OECD Reviews of Innovation Policy

Norway

OECD

ORGANISATION FOR ECONOMIC CO-OPERATION AND DEVELOPMENT

The OECD is a unique forum where the governments of 30 democracies work together to address the economic, social and environmental challenges of globalisation. The OECD is also at the forefront of efforts to understand and to help governments respond to new developments and concerns, such as corporate governance, the information economy and the challenges of an ageing population. The Organisation provides a setting where governments can compare policy experiences, seek answers to common problems, identify good practice and work to co-ordinate domestic and international policies.

The OECD member countries are: Australia, Austria, Belgium, Canada, the Czech Republic, Denmark, Finland, France, Germany, Greece, Hungary, Iceland, Ireland, Italy, Japan, Korea, Luxembourg, Mexico, the Netherlands, New Zealand, Norway, Poland, Portugal, the Slovak Republic, Spain, Sweden, Switzerland, Turkey, the United Kingdom and the United States. The Commission of the European Communities takes part in the work of the OECD.

OECD Publishing disseminates widely the results of the Organisation's statistics gathering and research on economic, social and environmental issues, as well as the conventions, guidelines and standards agreed by its members.

Corrigenda to OECD publications may be found on line at: *www.oecd.org/publishing/corrigenda*.

Foreword

This review of Norway's Innovation Policy is part of a series of OECD country reviews of innovation policy.* The review was requested by the Norwegian authorities, represented by the Ministry of Trade and Industry, and was carried out by the OECD Directorate for Science, Technology and Industry (DSTI) under the auspices of the Committee for Scientific and Technological Policy (CSTP).

The review draws on a background report prepared by the Norwegian authorities and on the results of a series of interviews with major stakeholders in Norway's innovation system as well as a peer review meeting within the CSTP's Working Party on Technology and Innovation Policy.** The review was drafted by Gernot Hutschenreiter (Country Review Unit, DSTI, OECD), Erik Arnold and Nick Vanston (consultants to the OECD), with contributions from and under the supervision of Jean Guinet (Head, Country Review Unit, DSTI, OECD).

This review owes much to Norwegian government officials at the Ministry of Trade and Industry who assisted in providing background information. The participants in the interviews in Norway as well as other individuals who supported the OECD team throughout the review process were also of great help.

* See www.oecd.org/sti/innovation/reviews

** During this meeting the examiners from OECD Member countries were Tricia Berman (Australia) and Sandra Bulli (United Kingdom).

Table of Contents

OVERALL ASSESSMENT AND RECOMMENDATIONS

Norway's economic performance has been consistently very good for a long time, and average real incomes are now among the highest in the world. The growing size and profitability of the offshore hydrocarbons sector has been a major factor, but even if it is excluded from the calculations, per capita GDP in mainland Norway is comparable to that of neighbouring Finland and higher than that of the major EU countries. Norway is also one of the best-performing countries in terms of growth and level of labour productivity, especially in private services.

However, the "summary innovation index" (SII), a synthetic indicator used in the EU's "Innovation Scoreboard", puts Norway below the EU25 average in 2007 (and the EU25 average is well below the US and Japanese scores). In addition, Norway's performance on this synthetic indicator has deteriorated over the years. Against this background, the "Norwegian puzzle" – *i.e.* that Norway "underperforms" against conventional S&T and innovation indicators despite its persistently high economic performance – has received some attention. However, it is well known that contributions to innovation and economic performance include forces such as, *inter alia*, the strong "social contract" between the state, labour and capital that promotes social welfare, and a high level of acceptance of technological change in the labour force. Low business sector R&D expenditure today can be largely "understood" by the industrial structure's smaller share of R&D-intensive industries than the OECD average. Non-R&D-based innovation, such as innovation in the service sector and in the organisation and the business model of enterprises, which is difficult to capture by available quantitative indicators, seems to underlie the exceptional productivity performance of the private services sector, which would otherwise be hard to explain. The key strategic task ahead is to maintain high, sustainable growth even after oil and gas production has peaked. Any foreseeable restructuring of the Norwegian economy compatible with this goal will entail a shift towards other knowledge-based activities. Policies to strengthen innovation capabilities, including the R&D component of the innovation system, are needed.

While this review argues that improved innovation capabilities require sustained increased investment in R&D, it also emphasises that Norwegian policy must translate these needs into concrete, mobilising and credible goals for all stakeholders. The "Barcelona objective" of 3% aggregate R&D intensity does not fulfil all of these criteria. In particular, given the nature of the Norwegian economy and its specialisation patterns, the likely failure to achieve this quantitative target could unfairly damage the credibility of Norway's science, technology and innovation policy.

A preferable approach might involve developing a (set of) sufficiently large programme(s) which could build on Norway's comparative economic advantages and capabilities in science and technology, and mobilise public and private actors towards common goals supported by a broad social consensus. More than many other countries, Norway has nurtured strong social support for action to contribute to solving problems of global relevance, such as sustainable development[1] and related issues. Large-scale programmes to address such topics could potentially have widespread impact on Norwegian industries and science and technology fields. Carefully crafted, they would strengthen the shift towards a more knowledge-based economy.

While framework conditions for increasing R&D and innovation are largely in place – especially those relating to the overall education and skill levels of the population – some changes in the governance of the innovation system seem necessary to facilitate prioritisation and efficient delivery of co-ordinated policies.

1. Sustainable development is a governing principle for the Norwegian government's domestic and foreign policy, and it is committed to pursuing trade and environmental policies that are mutually supportive.

Main strengths and weaknesses of Norway's innovation system

It is important to take specific aspects of the country's geography, economic specialisation patterns and cultural and institutional characteristics into account when assessing the state and potential of Norway's innovation system:

- Norway's topography is an economic asset, *e.g.* for developing international shipping, hydroelectric power, and more recently aquaculture, tourism, etc., but it is challenging in several respects (*e.g.* transport infrastructure, relatively isolated communities, scarce arable land).
- The development of the Norwegian economy has been shaped by the exploitation of natural resources. A long tradition in fishing has recently been complemented by a strong export-oriented aquaculture industry. The discovery and extraction of oil and gas, including the development of related industrial activities in engineering and services, have strongly affected the economy and have had a profound impact on the country's innovation and R&D system.
- Norway shares many cultural features with the other Nordic countries, including an egalitarian society, a high degree of individualism, and relatively high tolerance for uncertainty. These characteristics, on balance, seem to be conducive to innovation on the shop floor. Management tends to be consensus-oriented while individuals are expected to take responsibility and, in turn, resist being micromanaged.
- Nordic countries also share to some degree an organisation of the labour market which has become internationally known in its Danish version ("flexicurity"). This consists of a combination of a flexible labour market, participation of social partners in designing policy, generous arrangements for safeguarding the standards of living of those unable to work, and an active labour market policy focused on strengthening the competencies of the unemployed. In addition, Nordic countries have small wage disparities.

Analysis of strengths, weaknesses, opportunities and threats

Main strengths

- *Competitive natural-resource-based sectors, most importantly oil and gas.* Their demand for specialised goods and services provides opportunities for knowledge-intensive/value added businesses. In contrast to some other resource-based economies, Norway has made good use of this potential, notably by requiring foreign investors in the early development of the petroleum sector (before 1993) to invest in R&D and thereby to reinforce local technological learning.
- *A dynamic, high-performing private services sector.* There are many examples of innovative business models in many types of services, from telecommunications to media to retail trade as well as in services related to resource-based industries.
- *Disciplined and forward-looking economic policy.* Norway has ensured that oil revenues do not destabilise the domestic economy and that they will be available to meet long-term social needs.
- *Sound macroeconomic management and competition policy.* The economy is reasonably stable at the macroeconomic level. Competition policies have been tightened and are now on the same footing as in EU countries, so firms have incentives to "innovate their way out" of market pressures if policies are rigorously applied.
- *A highly educated labour force* (including scientific and technical skills) as concerns most age groups and both men and women strongly supports productivity and the ability to innovate; yet potentially serious challenges appearing in the OECD PISA 2006 results suggest that efforts must be made to secure the right quantity and quality of skilled personnel in the long term.
- *Strong consensus on the desirability of technological change and productivity increase generated by co-operation between the social partners.* This key social asset has helped Norway to build strong positions through technological modernisation and innovation in a number of traditional industries.
- *A labour market with sufficient flexibility for introducing new processes and products without too much disruption.* There seems to be less ingrained resistance to change in workplace conditions than in other OECD countries and an acceptance that there is no progress without change and that innovation benefits society as a whole, not just the innovating enterprise. It is not particularly difficult or expensive to lay

off workers for economic reasons, and the unemployed – including older people who wish to continue working – normally find new jobs relatively quickly. Undesirable aspects of the labour market are the very high incidence of sickness absence and disability pensioners.

- *Political commitment and institutional capabilities to foster science, technology and innovation. Fostering innovation has been a priority of successive governments.* Well-functioning institutions encourage innovation via information flows and both direct and indirect financial support. All levels are aware of the importance of innovation for economic performance and competitiveness. Norwegian policy makers have over the years been very active in developing a broad portfolio of support instruments for S&T and innovation. An important mission of Norway's research institutes has been to support industrial development through applied research. Policy formulation and delivery benefit from rich national sources of strategic intelligence on the research and innovation system.

Main weaknesses

- *A comparatively low level of R&D/innovation in some parts of the Norwegian business sector, especially in manufacturing.* This is a cause for some concern not because it reflects backwardness (in international comparisons Norwegian industries often perform acceptable levels of R&D) but because it reflects the need to restructure towards more knowledge-intensive industries while building on strength in existing ones. Increased R&D intensity in existing industries can also increase the competitiveness of the industries concerned, spillovers to other domestic sectors, the size of the market for knowledge generated by public research and the absorptive capacity of the Norwegian economy. Once oil and gas revenues peak, other exports will increasingly be needed to finance imports.

- In a rapidly globalising world, *Norwegian industry does not profit enough from R&D conducted abroad* and needs to adopt a more international perspective. Given its level of economic development and human capital, Norway does not attract enough R&D investment from abroad.

- *As in most other OECD countries, students and potential students are relatively uninterested in mathematics, science and technology courses.* The problem has been identified and measures to tackle it are in place.

Threats and opportunities

- *Failure to diversify*, in terms of energy sources and industrial structure, is widely recognised as a significant threat to future welfare, in light of the inevitable depletion of the oil and gas reserves and the demographic trend towards ageing. Government research and innovation policy, building on dynamic entrepreneurship in certain fields, actively promotes the development of new energy sources and industries. The risk of being locked into established industries, at the expense of new ones, should not be underestimated.
- *Policy contradictions may result in ineffectiveness.* An obvious example is the conflict between the centripetal forces needed to build critical mass and strong capabilities in many fields of research and the centrifugal forces of regional policy. Policy mechanisms to satisfy the need for both critical mass and regional empowerment are not in place.
- *A shortage of people with appropriate research skills.* While there is no fundamental shortage for the moment, there has been a fall in the numbers of students opting for scientific and technical disciplines. If the economy is to restructure in a more knowledge-intensive direction, the supply of people with scientific and technical skills must rise. Education policies have been reformed to address this challenge, but the results of these efforts are not yet clear.

Norway also has noteworthy opportunities:

- *Its current specialisation provides a strong base on which to develop and strengthen related economic activities.* A balance needs to be struck between policies to establish wholly new activities and those that build on existing strengths. Few new companies or industries arise out of nowhere; to emerge and grow they require customers, capabilities and ideas based on needs. Profitable industries and services can provide a springboard for the creation and growth of new related or unrelated activities by allocating the necessary resources provided that corporate governance (for firm diversification, spin-outs) and financial markets (for new technology-based firms) can play their role efficiently.
- *Norway's unique combination of capabilities and resources can be matched with global opportunities* to create and expand market niches, especially in areas in which global needs are pressing (*e.g.* clean energy, food, water, health, security, etc.). A dynamic, high-performing private services sector – which has received comparatively little policy attention so far – represents an important asset for developing such niches, which increasingly have characteristics of both manufacturing and services.

Strategic tasks and guiding principles

- A key strategic task for Norway is to prepare for "life after oil" even though advances in science and technology progressively push depletion of this source of income and wealth further into the future. A consensus seems to be emerging that structural change underpinned by strong innovative performance can help to maintain high, sustainable growth.
- Economic diversification requires a balanced strategy to develop existing knowledge infrastructure strengths and build new ones and to minimise institutional and infrastructural lock-ins to declining technologies and areas of knowledge.
- To be a successful competitor and partner in research and innovation, Norway must continue to raise the quality of Norwegian research. Best-practice research and innovation funding instruments are in place but they must be embedded in a governance framework which better safeguards against a recurring tendency towards fragmentation.

In accomplishing these tasks policy should be subject to key guiding principles:

- *A comprehensive approach to innovation.* Innovation policy should avoid an "R&D and high-tech myopia" and recognise the importance of non-technological innovation. Norway's strong resource-based sectors and services offer considerable scope for economic growth through the application of advanced science and technology. The "servicification" of manufacturing and the increasing technological component of services mean that both the manufacturing and services sectors need common capabilities to increase their knowledge intensity.
- *A systemic and evolutionary approach to the promotion of innovation.* A clear overarching strategy should inform policies that affect the dynamics and efficiency of innovation processes. Such policies should adapt to changes in the global environment and respond to the evolving needs of actors in innovation. They should help to improve the performance of the innovation system and sub-systems through continuous monitoring and assessment rather than define in advance an optimal innovation structure.
- *Competition and trust.* The increasing complexity, costs and risks involved in innovation enhance the value of networking and collaboration in partnerships between actors with complementary assets. This helps reduce moral hazard and transaction costs. Norway is well endowed with the necessary social capital to benefit from such co-operation and an appropriate competition policy framework.

- *Quality, relevance and critical mass.* Reconciling these objectives entails rigorous selection of research projects and teams eligible for public support, active involvement of research end-users in defining research priorities, and some concentration of resources in selected areas. An active regional policy should not lead to dispersion and/or duplication of research efforts.

- *Mobilising goals rather than quantitative targets.* The overriding objectives of science, technology and innovation policies should be formulated in terms of desired outcomes which can then be translated into resource requirements.

- *Market-friendly "clever" targeting.* Neither "picking winners" nor a pure bottom-up definition of policy objectives is the best way to use very limited resources. Some degree of top-down prioritisation is needed to focus efforts on areas in which national capabilities fit well with opportunities in national and global innovation networks. Market-friendly focusing devices include public-private partnerships for innovation.

- *Balanced internationalisation.* Most sources of the knowledge needed to sustain innovation-led growth must be "imported" from abroad in ways that already work quite well in Norway (FDI, labour mobility, cross-border licensing, etc.). They can also be accessed through outward investment and, more generally, active participation in innovation networks located abroad. There is scope for developing further the Norwegian innovation system's inward and outward linkages.

- *Good mix of public support instruments for R&D.* There is no known "correct" balance between tax incentives and grants for promoting R&D and innovation. Because both have advantages and disadvantages, offering both allows a wider range of actors to respond to a wider range of incentives than if only one or the other was available.

- *Advanced governance principles.* A clear distinction should be maintained between policy formulation and policy implementation. The latter draws on an effective mix of proven instruments: co-ordination, competition (*e.g.* competitive funding), co-operation (*e.g.* joint research projects); performance-based steering mechanisms (*e.g.* performance contracts, funding criteria). Changes in innovation policy governance should be embedded in Norway's proven system of disciplined and forward-looking economic policy.

Recommendations

Improving framework conditions for innovation

Existing framework conditions and policies are adequate for supporting a high level of innovation activity. However, there is scope for improvement in certain areas.

- *Identify obstacles to the growth of SMEs.* As in many OECD countries, small and medium-sized enterprises (SMEs) receive favourable treatment in terms of public support and employment regulations. Economic theory suggests there is a risk that this support may "crowd out" firms' own innovation priorities and activities, although there appears to be no evidence of this in Norwegian practice.

- *Correct mismatches in the demand and supply of skills.* Like most OECD countries, Norway needs to counteract a "flight from science" among young people. It has put in place a programme of action that appears to be coherent. However, its results are not yet clear. Monitoring and further efforts may be needed to increase the number of people trained in mathematics, science and technology to underpin a shift toward a more knowledge-intensive economic structure in future. Efforts to encourage students to study science and technology and to increase the supply of scientifically and technologically qualified teachers in schools should continue.

- *Achieve more balanced decentralisation.* A recent administrative reform *(Forvaltningsreform)* that delegates increasing budgetary authority to the regional level creates a significant challenge for developing overall research and innovation strategies that make sense at both the national and regional levels. Critical mass and consistency issues need to be addressed if Norway is to keep up with international trends (such as the concentration of effort implied by the European Research Area), given that its population is roughly equivalent to a single region of Germany or France.

- *Address regional disparities in access to venture capital without establishing too many regional funds.* When the government plays a smaller role in innovation-related venture capital – especially seed capital – its overall capacity to take risks is likely to be undermined. There should be a small number of national funds, such as Innovation Norway and SIVA, with access to regional distribution channels rather than many, small, locally controlled ones.

Improving governance of the innovation system

While innovation requires good framework conditions, OECD experience shows that effective public policy measures are needed to boost innovation performance. Overall, Norwegian science, technology and innovation policy corresponds to international good practice. Nevertheless there is room for improvement in some areas.

Strategic orientation, policy co-ordination and priority setting

- *Correct weaknesses in priority setting and governance in the public part of the research and innovation system.* Because of the strong sector principle in Norwegian governance many ministries micromanage policy implementation and under-exploit the capabilities of their agencies. This impedes overall priority setting for innovation in areas for which these ministries are responsible and prevents the design and implementation of the comprehensive innovation policies that are almost universally seen as desirable in Norway. It is important to complement the present sectoral principle with a stronger priority setting mechanism.
- *Consider ways of further strengthening overall policy co-ordination.* Setting broad overall priorities while exploiting the strengths of the sector principle for implementation may require further institutionalisation of active, consensus-oriented dialogue among the ministries and other stakeholders most concerned with innovation.
- *Tasking at least the Education and Research and Industry and Regional Development Ministries with co-authoring an innovation and research White Paper* instead of separate documents would be a further step towards policy co-ordination. The current drafting of a White Paper on innovation, a broad-based effort involving most ministries, is a step in the right direction.
- *Carry out a national-level foresight exercise* that could form a basis for building consensus among various stakeholder groups regarding areas on which to focus new efforts geared to Norway's potential and its societal ambitions. While Innovation Norway and the Research Council only recently began to use foresight techniques at low levels of aggregation following studies such as "Norway 2030" (carried out in 1998-2000), Norway has benefited from Parliamentary debate on national foresight and projection exercises at the aggregate level for many decades, and technology foresight would enrich this process and benefit from it.

- *Improve the understanding of the specific characteristics of Norway's innovation activities.* Closer study of parts of the Norwegian innovation system – such as complex, one-off North Sea projects – would lead to better understanding of these characteristics and would help define policy and select and interpret appropriate indicators.

Other governance issues

- *Provide more measures to encourage collaboration among universities, industry and the public research institutes.* The Norwegian institute system, especially its industrial institutes, is an important asset for innovation policy and performance.
- *Make greater use of demand-side innovation stimulation and support measures.* Norway should build upon the success of programmes like OFU/IFU to connect public procurement and innovation policies in ways consistent with national and international rules which make better use of private procurement needs.
- *Launch within the next one to two years a new round of evaluations of the three pillar agencies of research and innovation policy – the Research Council of Norway (RCN), Innovation Norway and SIVA* – to be carried out in the next two to three years to obtain a reasonably current view of the operations of these agencies which were last evaluated a decade ago.

Improving the effectiveness of research funding

The last few years have seen significant reforms in the funding of the knowledge infrastructure. University funding has moved towards good international practice, although the effectiveness of the new link between research quality and funding has yet to be tested. An interesting indicator-driven system for core funding of institutes has been proposed but not yet implemented. Norway has adapted good practice instruments, such as competence centres and centres of expertise, to complement existing instruments. These provide a wide range of opportunities for interaction.

Together, the Research Council of Norway, Innovation Norway and SIVA provide a comprehensive set of innovation instruments and services. However, the generally high level of earmarked funds is a major constraint on agencies' ability to play a flexible and creative role.

Against this background:

- *Consider the impact of earmarking.* The government's decision to "earmark" the Research and Innovation Fund effectively removes much of RCN's ability to act as an agent of change. Earmarking also means that RCN does not have the budget to act effectively as the source of strategic intelligence at this level in the system.
- *Take precautions so that the indicator-driven system to be implemented in the core funding of the institutes does not produce unintended effects,* such as excessively opportunistic behaviour, on an undesirable scale.

Promoting innovation in the business sector

- *Continue building technological capabilities in Norwegian industry.* The tax incentive aside, Norway spends modestly on direct encouragement of companies' performance of research and innovation (as opposed to paying for institutes to innovate on their behalf). Tekes-style industrial technology programmes and other measures that encourage companies to develop absorptive capacity or invest in innovation or R&D capacity will help put Norway on a trajectory towards more technology-intensive sectors while strengthening its positions in its traditional areas of strength.
- *Rebalance the policy focus.* At present, policies tend to focus too much on promoting R&D in SMEs and to recognise insufficiently the importance of larger companies as motors and vectors of change; for example, the Skattefunn scheme gives small firms strong incentives to undertake projects.
- *Consider additional measures to encourage R&D and innovation in Norway's sizeable and dynamic services sector.* The system of loans, grants and fiscal incentives for innovation should give equal treatment to the private services sector. This may mean changes in how requests for aid are vetted, given that it is often difficult to be precise about the expected gains in type or quality of service for proposed innovations in services.
- *Increase innovation in the public sector.* Given the size of the public sector, measures to promote innovation in this part of the economy are important potential drivers of national efficiency. Research and innovation efforts to this end would complement the suggested greater use of procurement as a tool for promoting innovation in industry.

- *Improve industry-science linkages by complementing RCN's instruments with measures that strengthen industrial absorptive capacity.* Differences among industries are important in this respect, so policy focus may need to be differentiated. There is a case for supporting the creation of R&D units by foreign multinationals in order to stimulate national learning in both business and academic sectors as well as for funding links between Norwegian companies and foreign (as well as Norwegian) academic institutions.

Maximising benefits from the internationalisation of R&D

Norway's industry has experienced the benefits of inward FDI for several decades, especially in the petroleum sector. The challenge now is to ensure that this learning experience can benefit as many sectors as possible, especially those such as services that can play a vital role to compensate for the future decline in oil-related activities. Another challenge is to ensure that Norwegian actors, especially SMEs, engage more actively in global innovation networks through outward FDI, mobility of personnel, cross-border co-operation, etc. The academic community is well placed but needs continued support to build critical mass in global networks.

- *Stress internationalisation in science, technology and innovation policy in all areas and provide greater support to international networking and partnerships, especially in industry.* This principle should affect both national funding mechanisms like RCN grants and participation in bilateral and multilateral arrangements.

- *Take a more strategic approach to international co-operation* with a view to improving the complementarity between national and international support programmes in science, technology and innovation.

- *Extend the scope of funding instruments beyond the EU.* While it is commendable that the Seventh Framework Programme aims to increase the participation of "third countries" in EU-funded research, this should be complemented by intensified Norwegian efforts to develop linkages with leading and emerging S&T powers outside the EU.

Summary table. Norway's NIS: strengths, weaknesses, opportunities and threats

Strengths	Opportunities
• Stable, high-performing economy underpinned by disciplined macroeconomic management • Generous and well-managed natural resource endowments • High level of social capital • High education and skill levels • Economic and socio-cultural framework conditions favourable to technical change • Political commitment to strengthen capabilities in science, technology and innovation • Strong industrial base, especially in industries conventionally classified as low or medium technology • Strong export-oriented resource-based industries • Innovative cluster around extraction, innovative services • Large and dynamic services sector • Segments of excellence in scientific research	• Still greater exploitation of value-added innovation in the resource-based industries • Build on sectoral strengths to create and enhance strengths in capital equipment and services and related industries, *e.g.* in alternative energy • Increased investment in research and innovation capabilities within existing industries • Further raise knowledge and innovation intensity of manufacturing and services • Develop new research and innovation strategies for the services sector • Exploit Norway's environmental advantages, *e.g.* through high-end tourism • Turn logistic constraints into innovation challenges • Create opportunities for unexpected benefits from breakthroughs in science and technology through sustained investment in excellent basic research
Weaknesses	**Threats**
• Low rate of R&D and innovation in manufacturing • Some weaknesses in university-industry links • Limited absorptive capacity in parts of industry • Mathematics, science and technology unattractive to students, potentially leading to skill shortages • Fragmentation/weaknesses in research and innovation governance and direction setting • Insufficiencies in some aspects of internationalisation (*e.g.* international learning of firms, R&D internationalisation in industry)	• Medium-term depletion of oil reserves, and long-term depletion of gas reserves • Failure to diversify and upgrade the economy towards more knowledge-based activities • Shortage of specialised human resources for innovation • Institutional lock-ins in research and innovation policy making and R&D performance • Fragmentation of needed concentrations of research and industrial strength through regional policy • Loss of credibility of science, technology and innovation policy

ÉVALUATION D'ENSEMBLE ET RECOMMANDATIONS

L'excellence des performances de l'économie norvégienne ne se dément pas depuis de nombreuses années, et le revenu réel moyen en Norvège se classe désormais parmi les plus élevés du monde. Si la taille et la rentabilité croissantes du secteur des installations pétrolières en mer ont certes joué un rôle déterminant dans cette évolution, le PIB par habitant en Norvège continentale reste comparable à celui de la Finlande voisine et supérieur aux grandes économies de l'UE, même lorsque les calculs excluent ce secteur. Les résultats de la Norvège en termes de croissance et de niveau de la productivité du travail figurent eux aussi dans le peloton de tête mondial, en particulier dans les services privés.

Cependant, l'« Indice de synthèse de l'innovation », un indicateur utilisé dans le Tableau de bord européen de l'innovation, classe la Norvège en deçà de la moyenne de l'UE25 en 2007 (cette dernière étant pour sa part largement inférieure aux résultats des États-Unis et du Japon). En outre, les résultats de la Norvège mesurés par cet indicateur synthétique se sont considérablement détériorés au fil des années. Ce constat a suscité des interrogations sur « énigme norvégienne » – qui réside dans le fait qu'en dépit de performances économiques durablement élevées, la Norvège obtient des résultats beaucoup moins satisfaisants au regard des indicateurs traditionnels de la S-T et de l'innovation. Ce paradoxe doit toutefois être relativisé si l'on considère que les parmi les facteurs contribuant de façon déterminante à l'innovation et la performance économique on compte entre autres l'existence d'un « contrat social » solide entre l'État, le monde du travail et le capital, propre à favoriser le bien-être social, et une forte propension, de la part de la population active, à accepter le changement technologique. La faiblesse actuelle des dépenses de R-D du secteur privé s'explique en grande partie par la structure industrielle du pays, qui compte moins de pans à forte intensité de R-D que la moyenne de l'OCDE. L'innovation ne dérivant pas directement de la R-D, comme celle qui concerne les services ou porte sur l'organisation des entreprises et les modèles d'entreprises, que saisissent mal les indicateurs quantitatifs disponibles, semble être à la base de la productivité exceptionnelle du secteur des services privés, performances qu'il serait difficile d'attribuer à d'autres causes. L'enjeu principal à l'avenir consistera à maintenir une croissance élevée et durable, et ce, même après l'apogée de la production pétrolière et gazière. Toute restructuration envisageable de l'économie norvégienne compatible avec cet objectif supposera une réorientation vers d'autres activités fondées sur le savoir. Des politiques de nature à renforcer les capacités d'innovation, y compris la composante R-D du système d'innovation, sont donc plus que jamais nécessaires

Si le présent Examen soutient l'idée selon laquelle l'amélioration des capacités d'innovation de la Norvège passe par un accroissement des investissements de R-D, il préconise fortement que l'action gouvernementale fasse en sorte de traduire cette nécessité en objectifs concrets, fédérateurs, crédibles et dont la réalisation associera l'ensemble des parties prenantes. L'Objectif de Barcelone, qui fixe à 3 % l'intensité globale de la R-D, ne répond pas à tous ces critères. Il est notamment vraisemblable que le pays, compte tenu des caractéristiques de son économie et de ses domaines de spécialisation, ne parvienne pas à atteindre cet objectif quantitatif, ce qui risquerait de porter indûment atteinte à la crédibilité de sa politique de la science, de la technologie et de l'innovation.

Il serait préférable de tabler sur un programme ou un ensemble de mesures suffisamment vaste pour mettre à profit les avantages économiques comparatifs et le potentiel du pays en science et en technologie, et mobiliser les acteurs publics et privés autour d'objectifs communs auxquels souscrit la société dans son ensemble. Plus que beaucoup d'autres pays, la Norvège a toujours misé sur une large adhésion de la société à l'action gouvernementale pour contribuer à résoudre des enjeux de pertinence mondiale tels que le développement durable[2] et des questions connexes. Des programmes de grande ampleur dans ces domaines pourraient avoir un impact potentiellement élevé dans les diverses industries et branches de la science et de la technologie norvégiennes. Bien concues, ils pourraient contribuer de façon décisive à accélérer la transition vers une économie davantage fondée sur le savoir.

Si les conditions cadres de l'accroissement de la R-D et de l'innovation sont dans une large mesure déjà en place, en particulier pour ce qui est des niveaux de formation et de compétences de la population, des changements doivent être apportés à la gouvernance du système d'innovation, afin de faciliter l'établissement des priorités et la mise en œuvre efficace de politiques coordonnées.

2. Le développement durable est un principe directeur de la politique intérieure et étrangère de la Norvège, qui se fait fort de mener des politiques commerciale et environnementale mutuellement compatibles avec cet objectif.

Les principales forces et faiblesses du système norvégien d'innovation

L'évaluation de l'état actuel et du potentiel du système norvégien d'innovation doit tenir compte de certaines spécificités de la Norvège relatives à sa géographie, ses types de spécialisation, sa culture et son cadre institutionnel.

- Si la topographie de la Norvège constitue un atout pour le développement de certaines activités économiques, notamment le transport maritime, la production hydroélectrique et, depuis plus récemment, l'aquaculture ou l'industrie touristique, elle est contraignante à d'autres égards (exigences en matière d'infrastructures de transport, isolement relatif de certaines communautés, relative rareté des terres arables, par exemple).
- L'économie norvégienne s'est construite sur l'exploitation des ressources naturelles. À la longue tradition halieutique du pays est venue récemment s'ajouter une aquaculture fortement orientée sur l'exportation. La découverte et l'extraction de pétrole et de gaz, ainsi que l'essor d'activités industrielles connexes dans l'ingénierie et de manière plus générale dans les services ont exercé une influence profonde sur l'économie. Cette mutation de l'activité économique s'est à son tour répercutée sur le système national d'innovation et de R-D.
- La Norvège partage avec les autres pays nordiques de nombreuses spécificités culturelles, dont le penchant égalitaire, un individualisme prononcé, et une tolérance à l'incertitude relativement élevée qui, au total, semblent favoriser l'innovation sur le terrain. De même, le style de gestion semble privilégier le consensus, et la demande implicite faite aux individus de travailler de manière autonome explique leur réticence à toute ingérence dans les détails de leurs activités.
- Les pays nordiques ont également tous en commun, dans une plus ou moins grande mesure, une organisation du marché du travail devenue célèbre dans le monde entier sous sa forme danoise (la « flexicurité »). Cette organisation réside dans l'association d'un marché du travail flexible, de la participation des partenaires sociaux à l'élaboration des politiques, de dispositifs généreux destinés à préserver le niveau de vie des individus qui se retrouvent dans l'incapacité de travailler, et d'une politique active du marché du travail, axée sur le renforcement des compétences des individus sans emploi. Les marchés du travail des pays nordiques se caractérisent par ailleurs par de relativement faibles écarts de salaires.

Analyse des forces, faiblesses, opportunités et menaces

Principales forces

- *La compétitivité des secteurs fondés sur les ressources naturelles, et avant tout sur le pétrole et le gaz.* La demande de ces secteurs en biens et services spécialisés représente des débouchés pour les entreprises à forte intensité de connaissances et à valeur ajoutée. Contrairement à plusieurs autres économies fondées sur les ressources, la Norvège est parvenue à tirer profit de ce potentiel, notamment en demandant aux étrangers investissant dans le développement initial (avant 1993) du secteur du pétrole d'investir également dans la R-D, ce qui a permis de renforcer l'apprentissage technologique local.
- *Un secteur des services privés dynamique et performant.* Les exemples de modèles d'entreprises innovants ne manquent pas dans de nombreux types de services, des télécommunications aux médias ou au commerce de détail, ainsi que dans les services liés aux industries fondées sur les ressources.
- *Une politique économique disciplinée et anticipative.* La Norvège a efficacement veillé à ce que les recettes pétrolières ne déstabilisent pas l'économie nationale, et à ce qu'elles restent disponibles pour répondre aux besoins sociaux à long terme.
- *Une gestion macroéconomique et une politique de la concurrence rationnelles.* L'économie connaît une stabilité relative au niveau général. La politique de la concurrence a été durcie et est désormais comparable à celles des pays de l'UE. Sa mise en œuvre rigoureuse encouragera les entreprises à innover pour réussir sur leurs marchés.
- *Des niveaux élevés de formation dans la population active* quel que soit l'âge, et pour les deux sexes (y compris pour ce qui est des compétences scientifiques et techniques), qui soutiennent vigoureusement la productivité et la capacité d'innovation bien que, si l'on en croit les résultats de la dernière enquête PISA de l'OCDE, des efforts demeurent nécessaires pour garantir sur le long terme la disponibilité d'un personnel qualifié adéquat sur les plans qualitatif et quantitatif.
- *Un consensus solide, fruit de la coopération entre les partenaires sociaux, autour du besoin de changement technologique et d'augmentation de la productivité.* Ce consensus est un atout qui s'est révélé essentiel pour asseoir, par la modernisation et l'innovation technologiques, les positions concurrentielles de la Norvège dans plusieurs industries traditionnelles.

- *Un marché du travail suffisamment flexible pour permettre une introduction assez harmonieuse des nouveaux processus et produits.* La réticence à l'égard du changement des conditions de travail semble moins farouche en Norvège, où la reconnaissance du lien entre progrès et changement, et du fait que l'innovation profite à la société dans son ensemble, et pas seulement à l'entreprise qui innove, semble l'emporter. Il n'est ni très compliqué ni particulièrement coûteux pour les entreprises de procéder à des licenciements économiques, et les personnes sans emploi – y compris les personnes âgées qui souhaitent continuer à travailler – trouvent de manière générale un nouvel emploi relativement rapidement. On peut toutefois relever d'autres caractéristiques du marché du travail qui sont moins flatteuses, tels que le fort absentéisme et l'importance des pensions d'invalidité.

- *Un engagement politique et des capacités institutionnelles pour promouvoir la science, la technologie et l'innovation.* La promotion de l'innovation a été la priorité des gouvernements successifs. Des institutions au fonctionnement bien rodé ont pour mission d'encourager l'innovation par la fourniture d'informations et un soutien financier direct et indirect. L'importance de l'innovation pour les performances et la compétitivité de l'économie est reconnue aux divers échelons de décision. Le gouvernement norvégien est depuis des années très actif dans la mise en œuvre de dispositifs de soutien de la S-T et de l'innovation, qu'ils soient originaux, ou inspirés de bonnes pratiques internationales, et leur gamme est par conséquent devenue vaste. Les instituts de recherche norvégiens, forts d'une mission déjà ancienne, ont accumulé des capacités élevées de soutien du développement industriel par la recherche appliquée. L'élaboration et la mise en œuvre des politiques tirent parti de l'existence de sources nationales abondantes d'intelligence stratégique sur le système de recherche et d'innovation.

Principales faiblesses

- *Le niveau comparativement faible de R-D/innovation dans certains pans du secteur norvégien des entreprises, en particulier dans certaines industries manufacturières.* Cette caractéristique est préoccupante – non pas tant parce qu'elle révèle un retard national (les industries norvégiennes affichent dans de nombreux cas des niveaux acceptables de R-D par rapport à d'autres pays) mais parce qu'elle témoigne de la nécessité de restructurer l'économie vers des industries à plus forte intensité de connaissances, tout en mettant à profit les atouts des industries existantes. Outre qu'il leur permettrait d'améliorer leur compétitivité, un accroissement de l'intensité de R-D de ces dernières se traduirait également par une augmentation des retombées positives dans les autres secteurs de l'économie nationale, ainsi que de la taille du marché des connaissances générées par la recherche publique et de la capacité d'absorption du savoir étranger par l'économie norvégienne. Les exportations de biens manufacturés seront de plus en plus nécessaires pour financer les importations, après que les recettes pétrolières et gazières aient atteint leur apogée.
- Dans un contexte de mondialisation rapide, *la capacité de l'industrie norvégienne de tirer parti d'une participation à des activités de R-D à l'étranger* est trop limitée et doit être renforcée par une intensification de l'internationalisation. Compte tenu de son niveau de développement économique et de capital humain, la Norvège ne bénéficie encore que trop peu des investissements de R-D de l'étranger.
- *Comme dans la plupart des autres pays de l'OCDE, on remarque un manque d'intérêt relatif de la part des étudiants ou élèves pour les mathématiques, les sciences et la technologie.* Ce problème est reconnu par les autorités qui ont lancé un train de mesures pour s'y attaquer.

Opportunités et menaces

- *L'incapacité à se diversifier,* que ce soit au niveau des sources énergétiques ou de la structure industrielle, est reconnue comme étant une menace considérable pour la prospérité future, compte tenu en particulier de l'épuisement inévitable des puits de pétrole et de gaz et du vieillissement démographique. La politique publique en matière de recherche et d'innovation s'appuie sur des entreprises dynamiques dans certains domaines pour encourager activement le développement de nouvelles sources d'énergie et industries. Mais il ne faut pas sous-estimer le risque d'une inertie structurelle qui favoriserait les industries en place aux dépens des nouvelles.

- *Les contradictions entre les différentes politiques risquent de se traduire par leur inefficacité croissante.* Un exemple évident est le conflit entre les forces centripètes nécessaires pour constituer une masse critique et des capacités élevées dans de nombreuses disciplines de recherche, et la force centrifuge de la politique régionale. Il n'existe pas de dispositifs institutionnels permettant de concilier les besoins de la masse critique et de la responsabilisation régionale.

- *Une pénurie de main d'œuvre possédant les compétences appropriées en recherche.* Si aucune véritable pénurie n'est encore à déplorer, on assiste depuis peu à une chute du nombre des élèves optant pour les disciplines scientifiques et techniques. La restructuration de l'économie en direction d'activités à plus forte intensité de connaissances nécessitera l'augmentation de l'offre de main d'œuvre possédant des compétences scientifiques et techniques. Les politiques de l'éducation ont été revues pour s'attaquer à ce problème. Les effets de ces réformes sont toutefois encore difficiles à évaluer.

La Norvège peut également saisir des opportunités importantes.

- *La structure de spécialisation de l'économie norvégienne continue de fournir une base solide de développement et de renforcement d'activités connexes.* Il convient par conséquent de doser efficacement les mesures visant à instaurer des activités entièrement nouvelles et celles destinées à tirer parti des atouts existants. Rares sont les entreprises ou les industries capables de se développer ex nihilo ; elles ont besoin de clients, de capacités et d'idées élaborées sur la base d'une bonne connaissance des besoins pour voir le jour et croître. Les industries et les services rentables existants représentent un tremplin solide pour la création et la croissance de nouvelles activités connexes ou non, par une allocation dynamique des ressources, à condition que le gouvernement d'entreprise (notamment par la diversification et l'essaimage) et les marchés de capitaux (les nouvelles entreprises à vocation technologique, par exemple) jouent leur rôle de manière efficace.

- *Il serait possible de faire concorder l'assortiment unique des capacités et des ressources de la Norvège avec des opportunités mondiales* pour créer et développer des créneaux, en particulier dans les domaines où les besoins mondiaux sont pressants (les énergies propres, les denrées alimentaires, l'eau, la santé, la sécurité, etc.). Le secteur des services privés, dynamique et performant – auquel les politiques publiques se sont jusqu'à présent relativement peu intéressées – représente un atout important pour le développement de ces créneaux, qui associent de plus en plus les caractéristiques de la fabrication et des services.

Missions stratégiques et principes directeurs

- Une tâche stratégique pour la Norvège est de préparer pour « l'après pétrole » – même si les progrès de la science et de la technologie repoussent sans cesse à une échéance plus lointaine l'épuisement de cette source de recettes et de richesse. Un consensus semble se dessiner autour de l'idée qu'un changement structurel sous-tendu par une forte innovation est dans cette perspective une condition nécessaire d'une croissance forte et durable.
- Diversifier l'économie requiert une stratégie qui concilie au mieux la consolidation des points forts existants des infrastructures du savoir et la création de nouveaux atouts, ainsi que des mesures permettant de contrer le tropisme institutionnel et infrastructurel à l'égard des technologies et domaines de connaissances en déclin.
- Il convient de poursuivre l'amélioration continue de la qualité de la recherche norvégienne pour permettre au pays de rester un concurrent et un partenaire performant dans les domaines de la recherche et de l'innovation. De ce point de vue, les dispositifs de financement existants sont conformes aux bonnes pratiques internationales, mais ils doivent être intégrés à un cadre de gouvernance qui permette de mieux contrecarrer la tendance récurrente à la fragmentation.

Les politiques destinées à mener à bien ces missions doivent être soumises à certains principes directeurs essentiels :

- *Une approche globale de l'innovation.* La politique de l'innovation doit éviter de se focaliser trop exclusivement sur la R-D et les activités à forte intensité technologique, reconnaître l'importance de l'innovation non technologique et admettre que les secteurs et les services liés à l'exploitation des ressources naturelles recèlent un potentiel de croissance important à réaliser grâce à l'application des sciences et des technologies avancées. Compte tenu du contenu croissant en services des activités manufacturières, de l'augmentation de la composante technologique dans le secteur des services, et de la forte compétitivité de la Norvège dans ce secteur, il est clair que de nombreuses ressources communes au secteur manufacturier et à celui des services conditionnent l'accroissement de l'intensité de connaissances dans les deux secteurs.

- *Une approche systémique et évolutive de la promotion de l'innovation.* Les politiques qui influent sur la dynamique et l'efficience des processus d'innovation doivent être cohérentes – et s'inscrire pour cela dans une stratégie globale – et pouvoir s'adapter à l'évolution de l'environnement mondial et aux besoins changeants des acteurs de l'innovation. Elles doivent viser à améliorer les performances du système d'innovation et de ses sous-systèmes sur la base d'un suivi et d'une évaluation permanents, au lieu de leur « prescrire » une structure optimale et « prédéterminée ».

- *La concurrence en confiance.* L'accroissement de la complexité, des coûts et des risques liés à l'innovation renforce l'intérêt de la constitution de réseaux et de la collaboration pour réduire les risques moraux et les coûts de transaction, et encourage donc une multitude de partenariats entre acteurs possédant des atouts complémentaires. La Norvège est dotée du capital social suffisant pour pratiquer la coopération et le partage des risques en matière d'innovation, ainsi que d'un cadre approprié de politique de la concurrence.

- *Qualité/pertinence/masse critique.* Afin de concilier ces trois objectifs, une sélection rigoureuse des projets et des équipes de recherche habilités à bénéficier d'un soutien public, la participation active des utilisateurs finals à la définition des priorités de la recherche, et une certaine concentration des ressources dans des domaines sélectionnés sont nécessaires. Une politique régionale active ne doit pas se traduire par la dispersion et/ou la duplication des efforts de recherche.

- *Des ambitions fédératrices plutôt que des objectifs quantitatifs.* Les objectifs principaux des politiques de la science, de la technologie et de l'innovation doivent être formulés en termes de résultats souhaités pouvant être traduits en exigences en matière de ressources investies, plutôt que l'inverse.

- *Un ciblage « intelligent » en harmonie avec le marché.* Une stratégie reposant sur le « choix des vainqueurs » *(« picking the winners »)* n'est pas viable, pas plus qu'une définition purement ascendante des objectifs des politiques, notamment dans les petits pays qui doivent utiliser au mieux leurs ressources très limitées. Un certain degré de définition descendante des priorités est nécessaire, afin de faire porter les efforts sur des domaines dans lesquels les capacités nationales peuvent répondre aux opportunités offertes par les réseaux nationaux et mondiaux d'innovation. Les partenariats public-privé pour l'innovation répondent à cette exigence de ciblage en harmonie avec le marché.

- *Une internationalisation équilibrée.* La plupart des sources de connaissances nécessaires à l'entretien d'une croissance tirée par l'innovation doivent être « importées » par le biais de divers vecteurs qui montrent en Norvège une bonne efficacité (IDE, mobilité de la main-d'œuvre, achats de licences, etc.). Elles sont toutefois également accessibles par le biais de l'investissement sortant et, de manière plus générale, par une participation active aux réseaux d'innovation situés à l'étranger. Il est possible et souhaitable d'étoffer davantage les liens bidirectionnels entre le système d'innovation de la Norvège et l'étranger.

- *Un dosage approprié d'instruments de soutien public à la R-D.* Il n'est pas possible de définir de manière normative l'équilibre optimal entre les incitations fiscales et les subventions destinées à promouvoir la R-D et l'innovation. Ces deux types d'instruments ont des avantages et des inconvénients spécifiques. En les mettant simultanément en œuvre, les pouvoirs publics stimulent un ensemble d'acteurs plus vaste et varié que s'ils avaient opté exclusivement pour l'un d'entre eux.

- *Des principes avancés de gouvernance.* Au plan institutionnel, une séparation nette doit être maintenue entre l'élaboration des politiques et leur mise en œuvre. Cette dernière doit reposer sur toute la gamme des mécanismes qui ont fait leur preuve : coordination, concurrence (financement concurrentiel, par exemple), coopération (projets de recherche conjoints) et dispositifs de pilotage axés sur les performances (contrats de performances, critères de financement, etc.) Tout changement de gouvernance de la politique d'innovation doit être compatible avec la politique économique disciplinée et anticipative de la Norvège.

Recommandations

Améliorer les conditions cadres de l'innovation

Comme indiqué précédemment, les conditions cadres et politiques liées permettent de soutenir un niveau élevé d'innovation. Des améliorations seraient toutefois possibles dans certains domaines.

- *Mieux identifier les obstacles à la croissance des PME.* Comme dans de nombreux pays de l'OCDE, les petites et moyennes entreprises (PME) bénéficient en Norvège d'un traitement favorable en termes de soutien public et de réglementations touchant à l'emploi. Théoriquement, ce traitement privilégié risque de rendre l'impératif d'innover moins pressant pour les entreprises concernées, même si rien n'indique qu'en pratique ce risque se soit déjà matérialisé en Norvège.

- *Remédier aux inadéquations entre la demande et l'offre de compétences.* À l'instar de la plupart des pays de l'OCDE, la Norvège doit combattre la désaffection des jeunes pour les sciences. Elle a lancé en la matière un programme d'action qui semble cohérent mais dont l'impact ne peut encore être évalué à ce stade. Il importe de suivre avec attention l'évolution de la situation et de maintenir l'effort pour augmenter l'offre de main d'œuvre possédant une formation en science et technologie, et pour former davantage de main d'œuvre hautement qualifiée que ce qu'exigent les besoins actuels, afin de soutenir l'évolution future vers une économie à plus forte intensité de connaissances. Les initiatives visant à inciter les élèves à entreprendre des études de science et de technologie et à accroître l'offre d'enseignants qualifiés dans ces disciplines doivent être maintenues.

- *Une décentralisation équilibrée.* La récente réforme de l'administration (« *Forvaltningsreform* ») qui, notamment, instaure une délégation croissante de l'autorité budgétaire à l'échelon régional pose des difficultés considérables pour l'élaboration de stratégies globales de recherche et d'innovation qui aient un sens aux niveaux national et régional. Les exigences de masse critique et de cohérence doivent être satisfaites pour permettre à la Norvège de s'adapter à l'évolution de l'environnement international (par exemple, la concentration d'efforts que suppose l'Espace européen de la recherche) – en gardant à l'esprit que sa population est à peu près égale à celle d'une région allemande ou française.

- *S'attaquer aux disparités régionales en matière d'accès au capital-risque.* Il convient toutefois d'éviter de multiplier les fonds régionaux. En effet, la fragmentation du soutien public au capital-risque – en particulier de capital de départ – pourrait se traduire par une diminution de la capacité du système de financement de prendre les risques de l'innovation. Plutôt que de soutenir une multitude de petits fonds gérés à l'échelon local, il conviendrait de s'appuyer sur un petit nombre de fonds nationaux tout en aménageant des accès appropriés aux canaux régionaux de distribution, comme Innovation Norvège et la SIVA.

Améliorer la gouvernance du système d'innovation

Si le maintien de bonnes conditions cadres pour l'innovation est nécessaire, l'expérience de l'OCDE démontre que des mesures publiques plus spécifiques sont nécessaires pour stimuler efficacement les performances nationales en matière d'innovation. A cet égard, la politique norvégienne de la science, de la technologie et de l'innovation est globalement conforme aux bonnes pratiques internationales, même si des changements sont désirables dans certains domaines.

Orientation stratégique, coordination de l'action et établissement des priorités

- *Remédier à certaines faiblesses du processus d'établissement des priorités et de la gouvernance de la composante publique du système de recherche et d'innovation.* La forte orientation sectorielle de la gouvernance en Norvège comporte le risque de voir de nombreux ministères s'impliquer par trop dans le détail de la mise en œuvre des politiques et sous-exploiter ainsi les capacités de leurs agences. Cette pratique n'encourage pas l'établissement de priorités dans des domaines de l'innovation dont les ministères sont responsables, pas plus que l'élaboration et la mise en œuvre de politiques compréhensives d'innovation dont le besoin est pourtant unanimement reconnu en Norvège. Il est par conséquent impératif de tempérer cette orientation sectorielle par un mécanisme d'établissement des priorités plus puissant et adaptable.

- *Rechercher les moyens de renforcer la coordination globale des politiques.* L'établissement de priorités générales et la mise à profit parallèle des atouts de l'orientation sectorielle dans la mise en œuvre peuvent nécessiter d'institutionnaliser encore davantage le dialogue entre les ministères et les autres acteurs concernés par l'innovation, dialogue qui est en Norvège très actif et axé sur le consensus.

- *Demander au ministère de l'Éducation et de la Recherche, et au ministère de l'Industrie et du Développement régional, au minimum, de rédiger en commun un Livre blanc sur l'innovation et la recherche* au lieu de poursuivre la production de documents séparés constituerait une étape supplémentaire vers la coordination des politiques. La rédaction en cours d'un Livre blanc sur l'innovation – fruit d'une vaste initiative associant la majorité des ministères – est un pas dans cette direction.

- *Entreprendre une activité nationale de prospective* dont les résultats pourraient permettre de dégager parmi les différents groupes d'acteurs une unité de vues concernant les domaines sur lesquels devront porter les nouvelles initiatives conçues spécialement pour répondre au potentiel et aux ambitions sociales de la Norvège. Si Innovation Norvège et le Conseil norvégien de la recherche a depuis peu recours à des techniques de prospective à un niveau relativement détaillé, suite à des études comme « Norvège 2030 » (réalisée en 1998-2000), la Norvège a bénéficié de débats parlementaires associés à des exercices de prévision et de prospective à un niveau agrégé depuis plusieurs décennies. Une activité nationale de prospective technologique pourrait enrichir ces débats tout en s'en nourrissant.

- *Améliorer la compréhension des spécificités de l'activité d'innovation en Norvège.* Une étude plus détaillée du fonctionnement de certains pans du système norvégien d'innovation – notamment des projets ponctuels complexes en Mer du Nord – renforcerait la compréhension des spécificités de la Norvège et faciliterait l'élaboration de politiques de soutien adaptées, ainsi que la sélection et l'interprétation des indicateurs.

Autres questions de gouvernance

- *Renforcer les mesures destinées à promouvoir la collaboration entre les universités, l'industrie et les instituts publics de recherche*, en prenant en considération que les instituts norvégiens, en particulier les instituts industriels, représentent un pilier fondamental du système d'innovation.
- *Recourir davantage à des mesures d'incitation et de soutien agissant du côté de la demande.* La Norvège doit tirer parti de la réussite de programmes tels que OFU-IFU pour établir un lien vertueux entre la politique des marchés publics et la politique d'innovation, selon des modalités conformes aux règles nationales et internationales, ainsi que pour tirer davantage profit des besoins des marchés privés.
- *Lancer sous 1 à 2 ans une nouvelle série d'évaluations des trois agences phares de la politique de recherche et d'innovation – le Conseil norvégien de la recherche, Innovation Norvège et la SIVA* – qui devra être menée au cours des 2-3 prochaines années afin de mettre à jour les connaissances relatives au fonctionnement de ces agences dont la dernière évaluation remonte à une décennie.

Améliorer l'efficacité du financement de la recherche

Le financement des infrastructures de la connaissance a connu ces dernières années de profondes réformes. Le financement des universités a évolué pour se rapprocher des bonnes pratiques internationales, même si le nouveau lien entre la qualité de la recherche et son financement n'a pas encore été évalué. À cet égard, un mécanisme nouveau pour le financement de base des instituts, reposant sur des indicateurs, a été proposé, sans toutefois avoir été mis en œuvre. La Norvège a complété la gamme de ses propres instruments en adaptant d'autres ayant fait leur preuve dans d'autres pays, tels que les centres de compétences et les centres d'expertise, afin d'étoffer, ouvrant ainsi la porte à de nombreuses opportunités d'interaction.

Les trois principales agences de soutien – le Conseil norvégien de la recherche, Innovation Norvège et la SIVA – apportent leur concours à l'innovation par un ensemble complet d'instruments et de services. Toutefois, le niveau globalement élevé de préaffectation des fonds – en particulier dans le cas du Conseil norvégien de la recherche – représente un obstacle majeur à la capacité des agences de jouer un rôle flexible et créatif.

Dans ce contexte :

- *Evaluer les répercussions de la préaffectation.* La décision des pouvoirs publics de « préaffecter » le Fonds pour la recherche et l'innovation réduit la capacité du Conseil norvégien de la recherche de jouer son rôle d'agent de changement. La préaffectation prive également le Conseil des ressources nécessaires pour fournir l'intelligence stratégique nécessaire à ce niveau du système.
- *Veiller à ce que le système fondé sur des indicateurs, qui doit être mis en œuvre dans le domaine du financement de base des instituts, ne produise pas d'effets indésirables*, tels que des comportements opportunistes et une chasse au financement trop systématiques.

Promouvoir l'innovation dans le secteur des entreprises

- *Continuer à s'attacher au renforcement des capacités technologiques de l'industrie norvégienne.* En dehors des incitations fiscales, la Norvège ne consacre que relativement peu de ressources à la promotion directe de la recherche et de l'innovation des entreprises elles-mêmes (par opposition au financement des instituts qui innovent pour le compte des entreprises). Les programmes technologiques industriels de style TEKES et d'autres mesures qui encouragent le développement des capacités d'absorption ou l'investissement dans l'innovation et la R-D des entreprises aideront la Norvège à initier son évolution en direction de secteurs à plus forte intensité technologique tout en renforçant ses positions dans les domaines traditionnels où elle enregistre déjà de bonnes performances.
- *Rééquilibrer les priorités stratégiques.* Les politiques sont pour l'heure exagérément axées sur la promotion de la R-D dans les PME et ne reconnaissent pas assez l'importance des entreprises de plus grande taille en tant que moteurs et vecteurs du changement. À titre d'illustration, le programme Skattefunn incite fortement les petites entreprises à se lancer dans des projets modestes de R-D.

- *Envisager la mise en place de mesures supplémentaires de promotion de la R-D et de l'innovation dans le dynamique et important secteur norvégien des services.* Le système de prêts, de subventions et d'incitations fiscales à l'innovation doit proposer un traitement équitable au secteur des services privés. Cela peut nécessiter des changements dans l'évaluation des demandes d'aide, compte tenu du fait qu'il est souvent difficile de quantifier avec précision les gains escomptés en matière de type ou de qualité de services, pour les innovations proposées dans ce secteur.

- *Intensifier les efforts portant sur l'innovation dans le secteur public.* Compte tenu de l'ampleur du secteur public, les mesures destinées à promouvoir l'innovation dans ce pan de l'économie sont susceptibles d'entraîner une amélioration de l'efficience au niveau national, et les efforts portant sur la recherche et l'innovation doivent par conséquent être renforcés. Ils viendraient en complément du recours accru suggéré aux marchés publics pour promouvoir l'innovation dans l'industrie.

- *Améliorer les liens entre l'industrie et la science en étoffant les instruments du Conseil norvégien de la recherche par de nouvelles mesures qui renforcent la capacité d'absorption de l'industrie.* Dans ce domaine, les disparités entre les industries sont importantes, et il serait peut-être nécessaire de différencier les objectifs des politiques. Il pourrait être utile de soutenir spécifiquement la création d'unités dédiées à la R-D dans les entreprises multinationales étrangères afin de stimuler l'apprentissage national dans les secteurs des entreprises et des universités, ainsi que d'appuyer financièrement le développement des relations entre les entreprises norvégiennes et les instituts universitaires étrangers (et norvégiens).

Valoriser les retombées positives de l'internationalisation de la R-D

L'industrie norvégienne bénéficie des retombées positives de l'IDE entrant depuis plusieurs décennies, en particulier dans le secteur du pétrole. L'enjeu consiste désormais à veiller à ce que l'expérience acquise soit mise au service du plus grand nombre possible de secteurs, en particulier de ceux qui sont susceptibles de compenser à l'avenir le déclin des activités liées au pétrole, par exemple les services. Un autre défi consiste à faire en sorte que les acteurs norvégiens, en particulier les PME, soient associés plus activement aux réseaux mondiaux d'innovation par tous les canaux possibles (IDE sortant, mobilité du personnel, coopération internationale, etc.). La sphère universitaire jouit d'une position favorable mais a besoin d'un soutien permanent pour constituer une masse critique au sein des réseaux mondiaux.

- *Mettre l'accent sur l'internationalisation dans la politique de la science, de la technologie et de l'innovation dans tous les domaines, et renforcer le soutien aux réseaux et aux partenariats internationaux, en particulier dans l'industrie.* Ce principe doit être mis en œuvre dans la gestion des dispositifs nationaux de financement, comme les subventions du Conseil norvégien de la recherche, comme dans celle de la participation à des coopérations bilatérales et multilatérales.

- *Adopter une approche plus stratégique en matière de coopération internationale* en vue d'améliorer la complémentarité entre les programmes de soutien nationaux et internationaux dans le domaine de la science, de la technologie et de l'innovation.

- *Élargir la portée des instruments de financement au delà de l'UE.* Si la Norvège peut se féliciter que le Septième programme-cadre vise à accroître la participation des « pays tiers » à la recherche financée par l'UE, elle doit aussi intensifier ses efforts pour développer ses liens avec des puissances en place et émergentes situées hors de l'UE, dans le domaine de la science et de la technologie.

Tableau synthétique. Le système national d'innovation de la Norvège : forces, faiblesses, opportunités et menaces

Forces	Opportunités
• Une économie stable et performante, qui bénéficie d'une gestion macroéconomique disciplinée • Des ressources naturelles abondantes et bien gérées • Un niveau élevé de capital social • Des niveaux élevés de formation et de compétences • Des conditions cadres économiques et socio-culturelles favorables au changement technique • Un engagement politique en faveur de la poursuite du renforcement des capacités dans la science, la technologie et l'innovation • Une assise industrielle solide, particulièrement dans les industries traditionnellement classées comme « à faible intensité technologique » ou « à moyenne intensité technologique » • La solidité des industries exploitant les ressources naturelles à vocation exportatrice • Une grappe d'innovation autour des services d'extraction et d'innovation • Un secteur des services important et dynamique • Des poches d'excellence dans la recherche scientifique	• Une exploitation renforcée de l'innovation à valeur ajoutée dans les industries fondées sur les ressources naturelles • Une mise à profit des atouts sectoriels pour créer et renforcer les points forts dans les biens d'équipement, les services associés et les industries connexes, comme les énergies alternatives • Une hausse des investissements dans les capacités de recherche et d'innovation au sein des industries existantes • Une augmentation de l'intensité de connaissances et d'innovation dans l'industrie manufacturière et les services • Le développement de nouvelles stratégies de recherche et d'innovation pour le secteur des services • La valorisation des atouts de la Norvège en matière d'environnement, par un tourisme haut de gamme par exemple • La transformation des contraintes logistiques en opportunités d'innovation • Tirer des avantages inattendus des découvertes scientifiques et technologiques fortuites grâce à des investissements continus dans une recherche fondamentale d'excellence
Faiblesses	**Menaces**
• Un faible niveau de R-D et d'innovation dans l'industrie manufacturière • Des faiblesses dans la relation entre l'université et l'industrie • Une capacité d'absorption limitée de certains pans de l'industrie • La désaffection des élèves pour les mathématiques, les sciences et la technologie, susceptible d'entraîner une pénurie de compétences • La fragmentation / les faiblesses de la gouvernance et de l'orientation stratégique de la recherche et de l'innovation • Des lacunes dans certains aspects de l'internationalisation (apprentissage international des entreprises, internationalisation de la R-D industrielle, par exemple)	• L'épuisement à moyen terme des réserves de pétrole et, à long terme de celles de gaz • L'incapacité à diversifier et à moderniser l'économie en direction d'activités à plus forte intensité de connaissances • La pénurie des ressources humaines spécialisées nécessaires à l'innovation • Un tropisme institutionnel en faveur des secteurs dominants dans l'élaboration des politiques de recherche et d'innovation et dans l'exécution de la R-D • La fragmentation des forces industrielles et de recherche induite par la politique régionale • La perte de crédibilité de la politique de la science, de la technologie et de l'innovation

Chapter 1

ECONOMIC PERFORMANCE AND FRAMEWORK CONDITIONS FOR INNOVATION

1.1. Introduction

This chapter first defines what is meant by economic performance and explores its linkages with innovation. In the medium term, and certainly in the longer term, there can be little or no increase in standards of living without constant innovation, which leads through various channels to productivity improvements. Norway's economic performance has been consistently very good on various measures. In contrast, on a number of measures, innovation input and output seem to be falling short of economic performance. This has prompted discussion of what is sometimes referred to as the "Norwegian puzzle". A number of explanations and factors that have been put forward in order to resolve this "paradox" are examined.

Next, Norway's current framework conditions for innovation are discussed. Overall, these are favourable and could probably support a higher level of innovation than is currently the case. Still there is scope for improving innovative performance by fine-tuning the framework conditions.

1.2. Innovation and economic performance

1.2.1. Measuring "economic performance"

A fundamental economic policy aim of governments is to create and maintain conditions that foster better economic performance, in the sense of rising material standards of living, at least in the long term. There may be different views on the appropriate role for governments in ensuring an equitable sharing of rising wealth among their populations and different approaches to dealing with the problems of pollution and social upheaval, for example, that often accompany rising prosperity, but there is general agreement that, other things being equal, rising incomes in real terms are to be preferred to stagnant or falling incomes, and the faster they rise the better.[3]

1.2.1.1. Gross domestic product

"Economic performance" is not a simple concept but one that covers several aspects of economic activity. It can refer to one or all of the following: total or per capita gross domestic product (GDP), real wages, productivity, competitiveness, trade balance, unemployment and inflation. In this section, the focus is on GDP (see Box 1.1), productivity, wages and competitiveness, because innovation has close links with these, but not directly with the other aspects of economic performance.[4]

Box 1.1. Measures of economic performance – GDP

"Rising material living standards" is nearly synonymous with rising real per capita GDP,[5] and the level of, and changes in, real per capita GDP are widely used measures of "economic performance". The reasons are both fundamental and practical. GDP is the aggregate value added of consumption and investment goods and services that are produced and sold by firms and persons operating within the country. In countries where goods and services are sold at prices determined by their markets, GDP is a reliable indicator of the value that society attaches at the margin to the resources used in their production.[6] Furthermore, the concept of GDP is well understood, and universally accepted methods for measuring it have existed for

3. The enjoyment of material goods and immaterial services is only one of the components of personal satisfaction, or happiness, or well-being. There is much evidence that beyond a certain point, increments to material living standards add less and less to well-being. Differences in self-reported levels of well-being across countries, or within countries, are much smaller than differences in real income levels. However, few people willingly accept a drop in their material living standards, and most feel that they would be happier if their incomes were higher. Hence, per capita GDP levels are an acceptable rough and ready indicator of levels of well-being. See for example, "Alternative Measures of Well-being" in OECD (2006), Chapter 6.

4. GDP *per capita* is as much an indicator of economic welfare as of economic performance, as it measures the possibilities for consumption (present and future, via investment) for every member of society, whether or not they are engaged in economic activity.

5. In countries that experience substantial changes in their terms of trade (the ratio of export to import prices) and whose foreign trade is large relative to GDP, the real disposable incomes of their populations might not move directly in step with GDP. The citizens of a country enjoying positive terms of trade gains can purchase more imports without having to produce more exports.

6. The existence of indirect taxes, goods and services that are not sold at market prices (*e.g.* many public services), inequalities in income distribution, and production of "bads" (*e.g.* pollution) complicate the picture. Nevertheless, valuing output at the prices that are set by markets (as opposed to central planners) remains the best available measure of economic welfare.

Box 1.1. Measures of economic performance – GDP *(continued)*

many years. The GDP of different countries during the same time period can thus be compared,[7] and the evolution of a given country's GDP over time can, for all practical purposes, be accurately measured. GDP is by definition a "value-added" concept: intermediate products used in subsequent stages of production are not counted in GDP.

When measuring economic performance, it is better to compare levels and growth rates of per capita GDP, rather than total GDP, which is an appropriate measure of a country's "economic power". For example, China is a much bigger player in the world economy than tiny Luxembourg, but real incomes in Luxembourg are very much higher than in China. It is not claimed that per capita GDP is a perfect measure of economic welfare, just that it is a widely available and reliable indicator. Incomes may be very unequally distributed across the population, and different countries have different demographic structures. Economic welfare, moreover, is only part of what constitutes overall welfare or happiness (see footnote 5).

One defect of per capita GDP as an indicator of economic performance is that it measures the value of goods and services produced within the country, not the value of goods and services that are purchased by residents, since some products are sold abroad (and some products are purchased from abroad). What counts for well-being is not the amount produced but the amount consumed. Foreign-owned firms transfer a part of their value added abroad as remitted profits. If the terms of trade are improving, a given quantity of exports (production) will allow the country to purchase more imports (consumption); its terms of trade are improving, and consumption can increase with no extra work effort. A part of the capital stock of plant and machinery is used up during the production process. As well as producing "goods", the production process unavoidably also produces "bads" in the form of pollution. In some countries, a significant proportion of output is produced illegally and escapes the statisticians. A better measure of economic welfare would take all of this into account, but no such ideal measure exists. It is possible to adjust GDP for changes in the terms of trade and the extent to which corporate profits are transferred abroad, and an allowance can be made for the using up of the capital stock. For most countries, however, such adjustments (which are often impossible to calculate accurately) make little difference to comparisons of growth of economic welfare over time, or across countries.

There are some exceptions, however. Ireland's per capita GDP is high, but much of the capital stock is foreign-owned and per capita gross *national* product, *i.e.* the part that accrues as income to residents, is typically 15 to 20% lower than per capita GDP. Norway's terms of trade are increasingly affected by the international price of its oil and gas exports and, as in the case of most OECD countries, its import prices have fallen both because of cheap imports from Asia, and because of falling prices of information and communication technology (ICT) equipment, of which Norway is an importer. Because of the rise in prices of oil and other important export commodities in recent years, and declines in some import prices, Norway's disposable income has risen faster than GDP.

7. To compare accurately the GDP of different countries during the same time period, their output must be measured using purchasing power parity (PPP) exchange rates. One difficulty is that it is not easy to calculate such exchange rates for countries whose economic structures and levels of development are very different. But it is not difficult to make reasonably accurate comparisons of the GDP levels of countries that are broadly similar, such as most OECD countries.

Table 1.1. Per capita GDP growth in selected countries

Country	Level, 2005*	1970s	1980s	1990s	2000-2006	1970-2006
Norway	156.9	3.71	1.97	2.85	1.80	2.67
Mainland Norway	116.8	3.60	1.54		2.34	
Denmark	121.9	1.91	1.90	1.89	1.35	1.81
Finland	117.1	3.33	2.63	1.11	2.79	2.43
Iceland	136.1	4.71	1.59	1.17	2.86	2.54
Sweden	120.6	1.47	1.87	1.15	2.30	1.63
France	108.7	2.95	1.29	1.11	0.91	1.63
Germany	105.7	2.63	1.65	1.34	0.95	1.72
Italy	104.5	2.91	2.15	1.19	0.53	1.82
United Kingdom	113.4	2.03	2.41	1.79	2.03	2.07
Ireland	136.8	3.04	1.77	5.44	3.40	3.41
United States	148.9	2.32	2.22	1.78	1.53	2.01
Japan	109.3	3.10	2.88	0.71	1.33	2.07
Australia	121.0	1.10	1.67	2.03	1.81	1.63
Chad	6.3		1.76	0.29	8.70	
South Africa	43.6		-0.60	-0.36	3.00	
China	24.4		7.40	8.50	9.10	
India	12.3		3.24	3.32	5.32	
Brazil	31.3		0.02	0.68	1.43	
Chile	43.9		1.00	4.54	2.95	

*Note: OECD=100. Levels are measured in terms of purchasing power parity (PPP). Growth rates are measured in constant-price national currency terms. Recent World Bank estimates would put the figure for China significantly lower.

Source: OECD, Statistics Norway, IMF.

Table 1.1 shows that Norway's economic performance in terms of per capita GDP growth has been consistently very good for a long time, and that average real incomes in Norway are among the highest in the world. The growing size and profitability of the offshore hydrocarbons sector has been a major factor, but even if it is excluded from the calculations, per capita GDP in mainland Norway is comparable to that in neighbouring Finland and higher than in the major EU countries. The table also shows the striking differences between levels of per capita GDP in most OECD countries and in developing countries, even comparatively prosperous ones. The data are expressed in purchasing power parity (PPP) to allow for valid comparisons.[8]

1.2.1.2. Labour productivity

Labour productivity is a closely related and widely used measure of economic performance. It is defined as the value of output produced per worker or per hour worked (see Box 1.2). Norway has recorded high productivity growth over the longer term. Between 1970 and 2005 labour productivity grew at an average rate of about 3% a year, in the same range as Finland and Japan. Among the countries listed in Table 1.2, only Ireland – which has undergone an extraordinary catch-up process – has realised higher productivity growth (of about one additional percentage point a year). Denmark and Sweden recorded significantly lower productivity gains than Norway.

Box 1.2. Measures of economic performance – labour productivity

Labour productivity is the value of output produced on average by each active member of the labour force. To allow for valid comparisons across countries and over time, productivity is measured as output per hour worked to take account of differences in standard working weeks, lengths of vacations, etc. Like GDP, productivity is measured in terms of value added rather than gross output. Productivity and its growth are more fundamental measures of economic performance than per capita GDP, because the latter is influenced by purely demographic changes. Table 1.3 presents data on labour productivity levels and changes for a number of countries. Norway is again one of the best-performing countries on this important measure.

Labour productivity and its growth are important because in the long term, it is virtually impossible for all members of society to enjoy rising living standards unless labour productivity also rises. In the short to medium term, the relationship is imprecise: the terms of trade may change, hours worked may change, and the proportion of the population actually working may also change. During much of the post-war period, per capita GDP in many OECD countries was boosted by a greater share of women in the workforce and the baby boom, which lifted the proportion of the population of working age from the early 1960s. This offset the trend towards shorter working hours and shorter working lives. At present, population ageing means that more workers retire each year than enter the labour force, while the trend towards shorter working hours continues. In some countries (Norway among them), there is little scope for further increases in female participation in the workforce. Therefore, the future evolution of per capita GDP will be largely determined by how labour productivity evolves.

8. Adjusting for purchasing power parity takes into account the fact that the prices of many non-traded services are lower in poorer countries. The PPP adjustment has the effect of narrowing the differences in GDP as between richer and poorer countries.

Table 1.2. Productivity growth in selected countries – total economy

Country	1970s	1980s	1990s	2000-05	1970-2005
Norway	4.33	2.29	2.70	2.66	3.03
Denmark	3.15	2.04	1.94	1.21	2.18
Finland	3.98	2.97	2.70	2.25	3.05
Iceland	4.96	1.21	-0.05	3.98	2.34
Sweden	2.29	1.25	1.94	2.58	1.94
France	3.64	3.19	1.82	1.84	2.71
Germany	3.72	2.05	2.51	1.51	2.55
Italy	3.97	2.25	1.44	0.39	2.18
United Kingdom	2.67	1.86	2.29	2.16	2.25
Ireland	4.64	3.64	4.33	3.11	4.01
United States	1.59	1.28	1.6	2.45	1.65
Japan	4.15	3.07	2.40	2.31	3.06
Australia	1.45	1.21	2.38	1.22	1.60

Source: OECD, Statistics Norway.

Productivity growth in Norway's manufacturing sector has been no faster than productivity growth in the rest of the economy (Table 1.3). This is unusual since productivity generally grows faster in manufacturing than elsewhere owing to greater opportunities for replacing labour by machinery and taking advantage of economies of scale. Labour productivity growth in Norway's services sector has been strong. As in other countries, productivity growth was rapid in the 1990s, as regulatory reform and technological progress in the financial, telecommunications and retail sectors spurred more efficient provision of services. However, Norway's performance has been consistently good for several decades, and indeed it has been a top performer. This is all the more remarkable as services are typically regarded as sheltered from the pressure of international competition and a sector with limited opportunities for productivity gains. Table 1.3 shows how the ratio of labour productivity gains in manufacturing and services have evolved over time in selected countries. Norway is clearly a special case. It is the only country in the sample for which labour productivity growth in services has been both high and higher than in manufacturing over a long period. Only Japan comes close. It is also noticeable that since the turn of the century, labour productivity growth in Norway's manufacturing sector has accelerated sharply, perhaps spurred by firms' reactions to strong real wage growth during the boom period, and the strength of the Norwegian currency.

Table 1.3a. Labour productivity – manufacturing

Country	1970s	1980s	1990s	2000-05	1970-2005
Norway	3.0	2.0	0.4	4.5	2.3
Denmark	5.4	1.8	1.9	2.8	3.2
Finland	4.7	4.4	5.2	5.4	5.3
Sweden	1.8	2.7	5.9	6.5	4.2
France	3.9	3.0	3.6	3.3	3.8
Germany	2.9	1.5	2.5	3.2	2.7
Italy	4.6	3.3	1.9	-0.7	2.9
United Kingdom	1.9	3.7	2.9	3.7	3.0
Ireland	5.3	5.8	6.4	6.6	6.3
United States	2.0	3.2	3.7	5.5	3.5
Japan	4.6	3.2	1.8	3.9	3.8
Australia	3.4	1.8	1.9		2.7*

Table 1.3b. Labour productivity – private services

Country	1970s	1980s	1990s	2000-05	1970-2005
Norway	4.3	1.3	2.8	3.2	3.1
Denmark	3.2	2.2	1.5	0.7	2.2
Finland	2.7	2.0	2.6	1.3	2.4
Sweden	1.8	1.1	2.0	2.5	1.9
France	2.6	2.0	0.8	1.0	1.9
Germany	2.2	0.9	1.8	1.2	1.7
Italy	1.8	0.0	1.0	-0.3	0.9
United Kingdom	0.4	1.7	2.4	2.4	1.6
Ireland	3.7	1.9	0.7	3.3	2.7
United States	0.8	1.0	2.0	2.0	1.4
Japan	3.7	3.1	2.1	2.6	3.3
Australia	1.0	0.3	2.3		1.4*

Table 1.3c. Labour productivity growth in private services as a percentage of labour productivity growth in manufacturing

Country	1970s	1980s	1990s	2000-05	1970-2005
Norway	145	63	636	70	136
Denmark	59	121	80	26	70
Finland	57	45	50	24	45
Sweden	97	40	33	39	46
France	67	65	23	30	50
Germany	76	61	71	37	64
Italy	40	1	54	49	31
United Kingdom	18	54	66	43	47
Ireland	71	32	11	51	43
United States		39	76	109	68
Japan	80	96	114	68	88
Australia	29	19	121		52*

*1970-2003.

1.2.1.3. Competitiveness

Governments and businessmen stress the importance of "competitiveness", a concept that covers several phenomena. For a country, it can mean that its foreign trade is balanced or in surplus, that foreign investors find it easy and profitable to start up a new business or acquire an existing one, that the legal and planning permission systems are predictable, transparent and efficient and not biased against foreigners, that the political and legal systems are predictable and stable, that corporate taxes and social charges are moderate and stable, that qualified and experienced personnel are available, and so on. For a discussion of various definitions of competitiveness, see Aiginger (2006) and Siggel (2006). For an individual firm, competitiveness can mean that it makes a good profit,[9] increases its market share and has a good reputation for reliability and innovative design. None of these factors is easy to quantify or measure unambiguously and many are subjective. Most importantly, competitiveness is a relative concept: a firm or an entire industry can be improving its efficiency, reducing costs, innovating, etc., but if rivals are improving their performance even faster, then competitiveness is deteriorating.

Competitiveness is clearly related to productivity: in a country, sector or firm with low labour productivity, that country, sector or firm will be uncompetitive, other things being equal. Labour costs are a very important aspect of those "other things". Despite very high labour costs, the most advanced OECD countries attract disproportionate amounts of foreign direct investment (FDI) because their labour productivity is very high and their legal and political systems are predictable.

Relative unit labour costs (RULCs), *i.e.* the ratio of labour costs to productivity across countries, are a standard measure of competitiveness.[10] It shows how a given country's relative competitiveness is changing over time. It does not show how or whether a country is more competitive in absolute terms than its trading partners. A country can be more competitive than its trading partners even if the RULC measure shows that its competitiveness is deteriorating, or vice versa. The standard measure is calculated for relative unit labour costs in manufacturing (the OECD used to calculate RULCs for

9 A firm may be profitable because it is "competitive" in the sense that it produces more efficiently than its competitors, in a market with no barriers to entry where consumers have a large variety of choice. A firm can also be profitable because it has no competition.

10 Unit labour costs, ULCs, are calculated as an index of how labour costs vary relative to productivity. It is not straightforward to compare "absolute" ULCs across countries, since the output mix varies and product prices differ. Hence the conventional measure is the ratio of an index of ULCs in a given country relative to an appropriately weighted set of ULC indices in trading partners.

the entire business sector but stopped owing to unreliable data). The indicator gives useful information on a country's competitiveness in world trade in manufactured goods, which account for most of world trade. Figure 1.1 shows how Norway's RULCs in manufacturing and those of nearby countries have evolved in recent years (an upward movement implies deteriorating competitiveness).

Norway's relative competitiveness in terms of unit labour costs was comparatively stable during the 1970s and 1980s, but deteriorated during the 1990s to stabilise again around the turn of the century. Denmark's trajectory was similar. Part of the development in Norway may be explained by the fact that high-technology manufactures, for which prices and costs have tended to fall over the medium term, are under-represented in Norway, relative to many trading partners (see Figure 1.1, first panel). Hence the measure of RULC exaggerates the loss of Norwegian cost competitiveness (however, because their prices have fallen, demand for high-technology manufactures has risen faster than for the more traditional manufactures produced by Norwegian industry). When comparing Figure 1.1 with Tables 1.1 to 1.3, it can be argued that while Norway's economic performance in terms of per capita GDP or productivity have been very good, there has been an erosion of international competitiveness because upward movements in real wages have not been fully compensated by productivity growth.[11]

It is clear that the erosion of competitiveness in the manufacturing sector owes more to movements in real wages and productivity than to higher inflation rates. Figure 1.1 shows that over 35 years, Norway's consumer prices have fluctuated in a rather narrow range relative to those of neighbouring trading partners (and consumer prices in general have moved in narrower bands than unit labour costs). There is no long-run tendency towards domestic inflation rates that are higher or lower than those in other countries, adjusted for changes in nominal exchange rates. This is also true for Denmark. In Sweden and Finland, relative inflation rates have been much more erratic (though trending downwards). In general, relative consumer prices have been more stable since the early 1990s, presumably as a result of the move towards monetary union.

11. In comparison with movements of real wages and productivity in trading partners.

Figure 1.1. Measures of relative competitiveness[1]

Relative unit labour costs, manufacturing

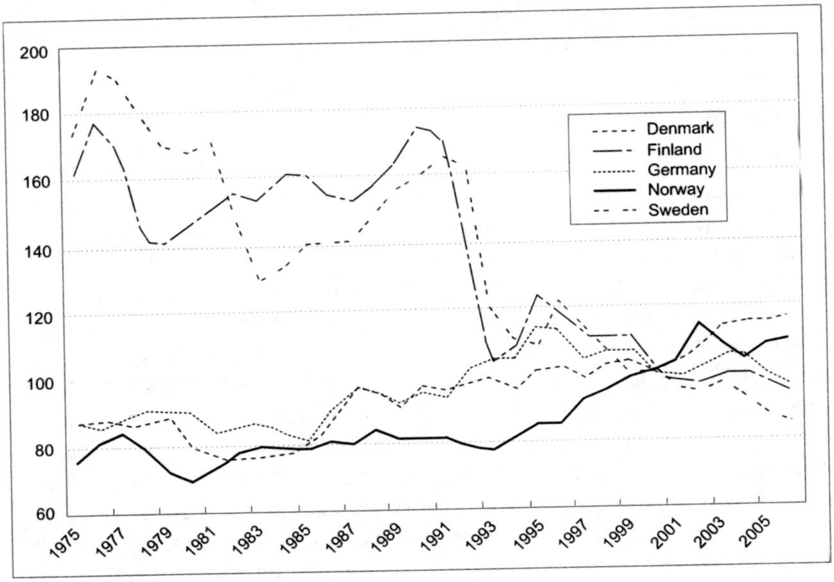

Relative consumer prices

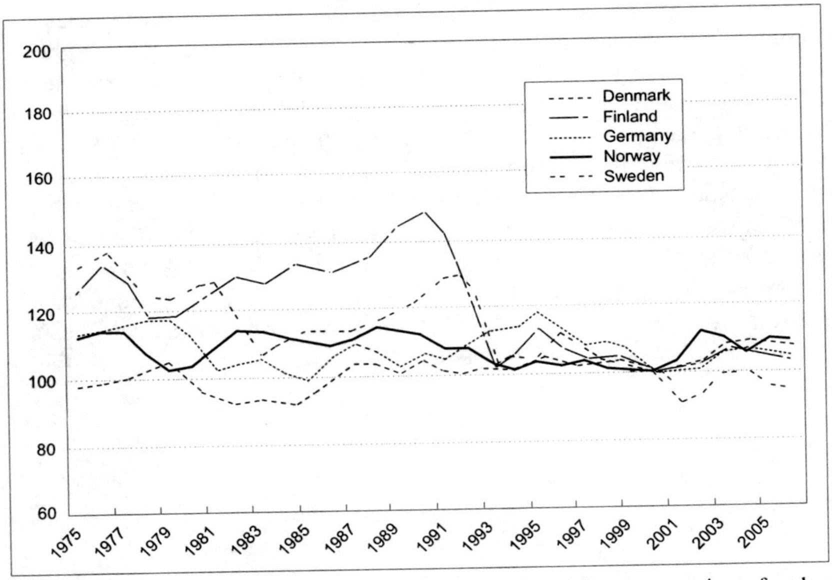

Note: In these graphs on relative competitiveness, the unit labour costs and consumer prices of each country are measured relative to a trade-weighted average of costs and prices in trading partner countries. For details of the calculations, see M. Durand, "Method of Calculating Effective Exchange Rates and Indicators of Competitiveness", OECD Economics Department Working Paper No. 29, at www.oecd.org/dataoecd/26/57/2345608.pdf.

The combination of deteriorating competitiveness in the manufacturing sector but comparatively stable relative inflation rates implies that to regain competitiveness in the longer term, real wage growth in Norway's manufacturing sector would have to be slower than that of productivity growth. The fact that profitability in the Norwegian manufacturing sector is currently high does not necessarily weaken this argument. Profitability in manufacturing has also been high in many other OECD countries in recent years, and wage shares low.

Box 1.3. Raising productivity

There are four main channels for increasing labour productivity: *i)* an increase in the ratio of capital equipment to labour input; *ii)* use of "better" capital equipment; *iii)* rising skill levels of workers via education and/or on-the-job training; and *iv)* more efficient use of existing capital equipment and skilled labour.

Higher capital intensity

In basic economic theory, a rising ratio of capital equipment per worker is typically portrayed as the main force behind rising labour productivity. A worker with a wheelbarrow is more productive than one without, and a worker with a bulldozer is more productive than one with a wheelbarrow. Historically, the industrial revolution was essentially a period in which machines powered by water, steam and then hydrocarbon fuels and electricity replaced working methods based mostly on human or animal muscle power. Productivity soared. Even in the most advanced industrialised countries, the process continues, with automation of many industrial processes, increasing use of specialised machinery for tasks formerly performed by hand and the rapid spread of information and communication technology (ICT). However, the productivity gains associated with rising capital/labour ratios become smaller and smaller as the ratio increases. In the long run, therefore, productivity growth will slow to a halt if workers are simply equipped with more and more of the same kinds of machines.

"Better" physical capital

This trap can be avoided by using "better" capital, *i.e.* machinery and equipment that incorporate new or improved technology which can be used to produce new products or to produce existing products more efficiently. For example, the arrival of electric motors in factories permitted much more flexible working methods than steam engines, boosting productivity even though the capital/labour ratio did not radically change. Today, labour productivity has risen in many sectors because of the wide use of cheap information and computing technology. Computers in the early 21st century and much of their software are far more powerful and far cheaper than those of a few decades ago, so that in value terms, the capital/labour ratio has actually fallen in areas that have used them intensively since the beginning. As with the introduction of electric motors in factories, the impact of ICT on productivity will be maximised when production methods and workplace habits are fully adapted to the possibilities offered by the technology. Technological progress of this kind seems limitless and is the major force behind continuing productivity growth in even the most advanced industrialised countries.

.../...

Box 1.3. Raising productivity *(continued)*

Better human capital

A common feature of advanced countries is the difficulty the unskilled face in finding and keeping jobs that pay enough to enjoy a decent standard of living. The logical explanation is that their productivity levels in any available job are too low to make them worth more than the legal minimum wage to an employer, and possibly not even that.[12] This is a major change from a century ago, when most workers were unskilled: little or no full-time education after age 13-15 and little or no subsequent training. While today's workers are not more intelligent than those of a few generations ago, there is no doubt that the extra years of full-time education they have received result in significantly higher productivity levels, even when education beyond the legal minimum is not in areas directly relevant to their jobs.

The question arises of whether increases in human capital are subject to the same law of diminishing marginal returns as physical capital. Analysis of the impact of education on productivity shows that a population of working age that is both literate and numerate, with 10-12 years of full-time education, results in very large economic returns to society. The costs to society are well below the returns. The returns to higher education (university or equivalent) are lower (and seem to benefit the individuals concerned, in the form of higher salaries, more than society as a whole). But human capital can be boosted by on-the-job training as well as formal education, and use of new technology often requires training the operatives. Hence the returns to increased or "better" human capital do not necessarily fall at the margin, and there is no evidence that societies' investment in education is not cost-effective at the margin.

More efficient working methods

An all-important source of continuing productivity growth is more efficient use of existing resources both to produce goods and services and to distribute and market them. Empirical analysis of individual firms in a given industry shows that there are large differences in productivity levels and profitability between the least and most productive firms, even allowing for differences in worker skills, types of capital equipment and location. This shows the importance of management skills in boosting productivity. Such productivity gains can be realised indefinitely, because as the business environment evolves the challenge of adapting will always exist. Almost by definition, there will always be enterprises that are operating below "best practice", and best practice itself is constantly evolving as new ideas, techniques and technology spread throughout the economy.

12. It is sometimes claimed that this is the result of a particular type of technological change that is biased in favour of skilled workers. But this cannot be the whole story. China and other Asian countries have created hundreds of millions of jobs for people who are unskilled by the usual definitions, where workers operate modern computer-controlled high-technology machinery to produce sophisticated goods at low cost.

1.2.2. Innovation and productivity

Innovation is regarded as an important, probably the most important, way to boost labour productivity in the longer term. There are two quite different kinds of links between innovation and productivity: via products and via processes and distribution.

Product innovation – the successful production of a new or improved good or service – does not in itself necessarily raise productivity. Of course, if a new product can be sold at a higher price than an old one using the same resources of capital and labour, the value added of the factors of production is higher and in that sense productivity is higher. However, the price of the new product is likely to drop over time as competitors enter the market, sometimes with even newer products. Productivity for that individual product, as conventionally measured, then falls. Nonetheless, product innovation allows productivity to keep rising in the longer term. In the absence of new products, consumers would eventually become satiated with existing products, and growth would grind to a halt. Indeed, most growth since the industrial revolution has come about because of the development of products that either did not previously exist – steam and electricity-powered engines for use in transport and manufacturing, aviation, photography, electronics – or have been enormously improved, *e.g.* pharmaceutical products. Successive innovations create new demands to match the new supplies and have therefore allowed productivity and incomes to continue to grow. This is a necessary channel for long-term growth.

Quantifying the overall, economy-wide impact of product innovation on productivity is not straightforward. Individual case studies can trace the history of a particular product innovation, such as a particularly successful new drug or computer tomography, but there is no reliable quantitative indicator for all of the successful (and unsuccessful) innovative products that enter the market each year or of the resources used (or wasted) in their production. It is clear that the greater the rate of innovation, the faster the rise in labour productivity through this channel over time. Productivity gains will be rapid in countries that adopt the innovations via licensing and leasing, as well as in those responsible for the original innovations. This helps explain part of the fast growth of productivity throughout the industrialised world in the post-war period when inventions made just before or during the war (antibiotics, jet engines, computers, television) were exploited worldwide.

Process and distribution innovation is another powerful way to raise labour productivity. It is also a very powerful independent source of growth, over and above the contributions of increases in labour and capital inputs. "Innovation" here is used in its broadest sense, encompassing all new or improved ways of producing and distributing a given set of goods and services. In some cases, techniques completely new to the firm may be involved, while in others, more efficient use of existing production methods and labour and capital resources lowers costs and raises productivity. Some improvements may simply be regarded by the firms concerned as adapting to best practice rather than as innovation. Some innovations will mean costly investment to apply and develop new or better production techniques. Others may merely involve a reorganisation of employee responsibilities to facilitate teamwork and ensure that the talents of individual employees are exploited optimally. All lead to higher output or lower costs, or both, *i.e.* to higher labour productivity. In quantitative terms, this type of innovation is essentially identical to multifactor productivity (MFP). MFP is the contribution to output over and above the contributions from increased inputs of labour and capital,[13] and is estimated to be the most important contribution to per capita GDP growth over the long term in OECD countries. It follows that the higher the rate of this type of innovation, the higher the rate of growth of MFP, and, other things being equal, the faster the growth of labour productivity. It also follows that if MFP growth is measured as being fast (or slow), the rate of innovation, broadly interpreted, in processing and distribution must also be fast (or slow), whatever the other indicators of innovation activity may show.

It can thus be concluded that in a country or a sector where labour productivity rises briskly and steadily over a long period and multifactor productivity also grows strongly, innovation activity must necessarily be strong. As shown above, Norway has indeed recorded high labour productivity growth over an extended period of time. The same can be said about MFP growth, which can be interpreted as measuring "technical progress" after other factors – such as quality improvements in physical and human capital (see Box 1.3) – have been taken into account.

13. Note that at least part of the contribution to growth that comes about via "better" capital equipment and "better" human capital will also appear as multifactor productivity. Adjusting for these factors tends to lower the estimated growth rates of MFP, but without eliminating them or greatly changing their ranking across countries. A well-known complication in measuring and comparing MFP across countries arises from differences in the assumptions made when measuring the capital input and the availability of detailed capital stock data. Access to free or low-cost natural resources also affects the calculation of MFP. Hence, small differences in the calculated growth of MFP across countries may not be significant, although large ones probably are.

Multifactor productivity growth was quite high in Norway in the 1990s (Table 1.4a). While it has slowed during the past decade it remains at the higher end of the OECD countries in the table, just behind Ireland and Finland. MFP growth has been particularly strong since the end of the 1990s in the private services sector, but has also picked up in manufacturing over the past ten years.

Table 1.4a. Multifactor productivity growth in selected countries

	1973-81	1985-88	1989-96	1997-2005
Norway	1.50	1.11[1]	2.70	2.12[2]
Denmark		1.13	1.60	0.00
Finland		2.47	1.85	2.44
Sweden		0.50	0.58	1.76
France		2.10	1.12	1.26
Germany				0.95
Italy		1.71	0.99	-0.10
United Kingdom		1.17	1.01	1.40
Ireland		2.30	3.69	3.37
United States		0.92	0.80	1.55
Japan		3.14	1.67	1.12
Australia		0.56	1.29	1.18

Table 1.4b. Mainland Norway: multifactor productivity growth by sector

	1973-81	1982-88	1989-96	1997-2006	2005-06
Business sector	1.5	1.1	2.7	2.1	2.6
Manufacturing	0.8	1.4	0.6	2	2.3
Private services	1.5	1.1	3.0	2.6	2.6

1. 1982-88.

2. 1997-2006.

Note: Calculation of MFP growth depends inter alia on estimates of the capital stock. The EU/OECD/UN/IMF/World Bank conventions for calculating capital stocks were modified in the late 1990s, but few countries have provided historical data on the new basis for before 1985. MFP data for earlier years are not directly comparable with those estimated on the new basis. Norway does not participate in the OECD capital-stock dependent part of the OECD Productivity Database.

Source: OECD, Statistics Norway.

To what extent does this finding agree with other indicators of innovation activity, and how does Norway fare by comparison with other countries?

1.2.3. A "Norwegian puzzle"?

Since about 2004 the so-called "Norwegian puzzle" has drawn much attention.[14] In essence, it is that despite weak innovation inputs and even weaker outputs (measured by a number of standard indicators), Norwegian per capita incomes are very high in international comparisons, even excluding the direct impact of hydrocarbon exports.

The "puzzle" is particularly visible when R&D-based indicators are used to benchmark the Norwegian innovation system.[15]

Some Nordic countries substantially increased their R&D spending in the past 10-15 years. In Norway, measured as a percentage of GDP, it is markedly lower than in other Nordic countries and somewhat below the OECD average.[16] The somewhat broader European Innovation Scoreboard (which shows Norway to be performing rather poorly) is discussed below.

Koch and Hauknes (2007) have summarised some of the comments on the puzzle that have been made:

- Innovation scoreboards, including the European Innovation Scoreboard, tend to place more emphasis on R&D-based indicators than is warranted by the actual role of R&D in determining economic performance.
- To some extent, Norwegian GDP is "inflated" by the country's large oil and gas export revenues, most of which are saved rather than spent. Hence its use as the denominator in calculating R&D intensity biases the result downward in comparisons with many other industrialised countries. However, even adjusting for this, Norwegian R&D spending remains lower than in other Nordic countries.
- Norway does not have much high-technology industry in the sense of large-scale industry that invests heavily in R&D and then spreads its sunk costs across the output it produces. In 2000, 40% of all business sector R&D spending in the United States was accounted for by the aerospace, ICT, pharmaceutical and instrument industries, all of which are major industries in that country. By contrast, for the same set of industries in Norway, where they are comparatively small, the figure

14. For several years, there was discussion of the (opposite) "Swedish paradox", namely, why, when the country performs so much research and tops various lists of science, technology and innovation indicators, has its level of per capital GDP obstinately stayed around the OECD average and even declined slowly over time relative to that benchmark.

15. The performance of Norway's R&D system in an international comparison is examined in more detail in Annex A.

16. In recent years, R&D as a percentage of nominal GDP has been biased downwards because of the high volumes and rising prices of oil and gas exports.

was less than 17%. Despite their lesser overall importance, however, Norwegian high-tech firms tend to be more research-intensive than their trading partners.

- Figure 1.2 shows that when adjusted for industrial structure, R&D spending in Norway is not very different from that in the largest OECD countries, although some distance behind the leaders. If all OECD countries had the same industry structure, Norwegian industry would be the fourth most R&D-intensive country in the OECD, instead of the tenth.

Figure 1.2. R&D intensity in the business sector adjusted for industrial structure[1]

Average over 1999-2002

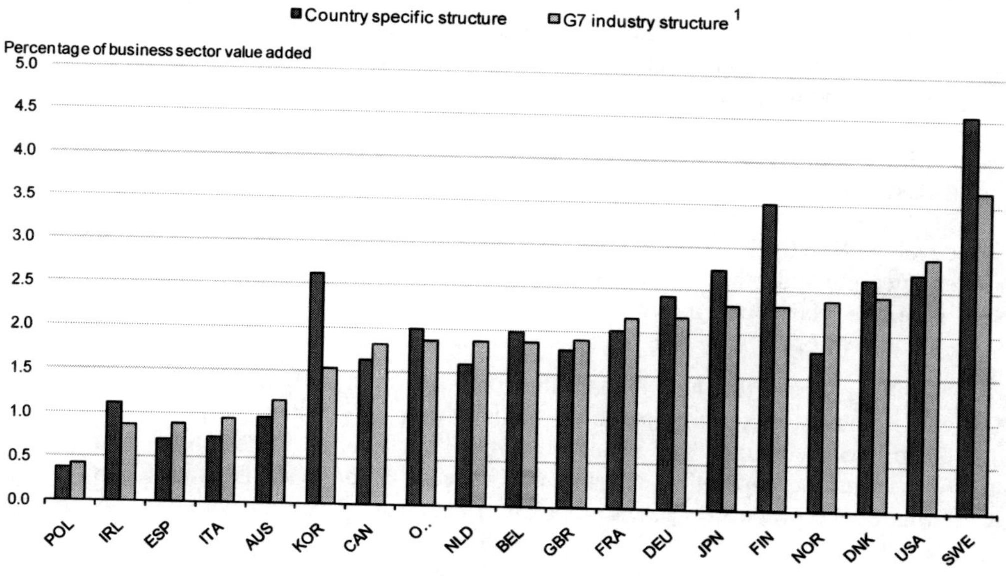

1. All countries are assumed to have the same industry structure. Calculated on the basis of R&D intensity per industry with the weights of each industry corresponding to its average share of total business sector value added across G7 countries.

Source: OECD ANBERD and STAN databases.

- Much of Norway's knowledge-intensive engineering activities (such as the major one-off oil and gas installations on and under the North Sea), actually involve substantial development efforts, work that in large manufacturing enterprises is likely to be counted as R&D. Another view is that R&D matters most for high-technology industries but that the best productivity growth is achievable in low-technology industries with minimal spending.

- Disciplined macroeconomic policy combined with the openness of the economy and the specific socio-cultural framework have been major "non-technological" contributors to strong economic performance.

The points made above are valid ones, but they are not reasons for complacency. Most other available indicators of innovation, which are not subject to the same kinds of criticism, also imply a comparatively low level of innovation activity in Norwegian industry, especially, but not only, compared with its Nordic neighbours. Norway is a rich country, but it would be even richer if innovation activity were more intense, as it could be, given that framework conditions for innovation activity (see next section) are relatively favourable.

As noted above, Norway does not possess many large firms in sectors with particularly high R&D intensities. It does not in itself explain why Norway's industrial structure is unusual. In practice, most Norwegian business firms – particularly large firms in the core national clusters of petroleum, maritime and marine industries – are not seeking to create new science-driven products and industries. Instead, their innovation strategies often focus on more efficient processes based on natural resources. Norway has developed sophisticated supplier industries, for example, in the machine and engineering industry, as well as highly sophisticated biotechnologies that serve the petroleum industry, shipping, marine aquaculture and fishing (Remøe *et al.*, 2004, p. 88).

Other factors that help explain the combination of high productivity and weak innovation activity, as conventionally measured, should probably be taken into account.[17] They include the very high level of education of most of the working population. Norway ranks second among OECD countries in terms of the proportion of the population between 25 and 64 with Type A (lengthy, theory-based) tertiary qualifications (30% compared to the OECD average of 19%). For the age group 25-34 this proportion is 29%, compared to an OECD average of 24% (OECD 2007c). Norwegians spend more time in formal education than people in other OECD countries: 13.8 years, compared with a mean of 12 years (OECD, 2007). As a result, the workforce has high capabilities, and the performance of both routine activities and quality/improvement-focused activities such as continuous improvement (which improve performance but tend not to count either as R&D or as innovation) also tend to be of high quality. The contribution of the work-force's high average level of education and competence to productivity,

17. Some of these factors are discussed in section 1.3 on framework conditions for innovation.

innovation and entrepreneurship is not well explored in the literature and is probably underrated.

1.2.4. The example of the Innovation Scoreboard

As emphasised above, quantitative measures of innovation activity such as R&D spending and numbers of patents granted have limitations because they do not capture all innovative activities and because it is not possible to distinguish between commercially successful innovations and others. A different and complementary insight into the innovation process can be gained from survey evidence. The EU has performed surveys in this area for some years, and the Community Innovation Survey (CIS) includes Norway in the sample.

Figure 1.3. Evolution of innovation activities in European countries

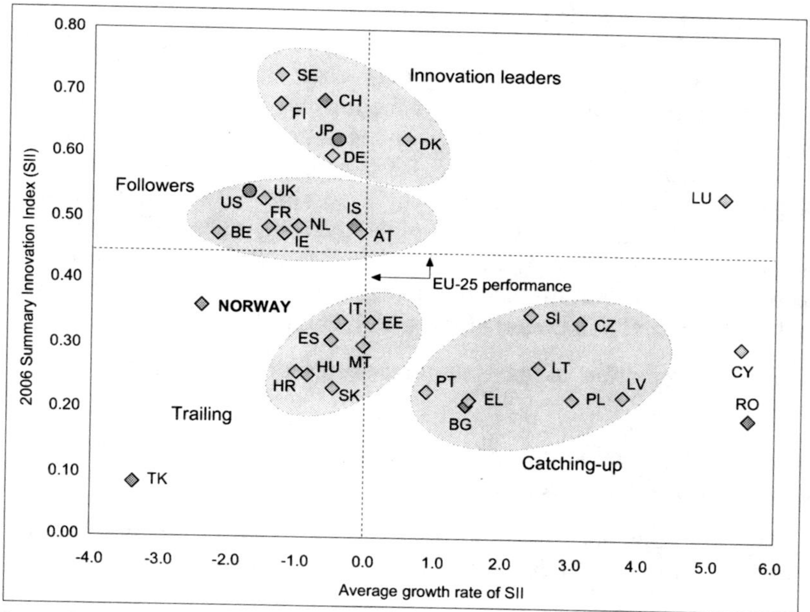

Source: European Innovation Scoreboard 2006.

CIS data are also used as one input into the EU's annual *European Innovation Scoreboard*. Their synthetic indicator, the summary innovation index (SII) puts Norway below the EU25 average in 2006 (and the EU25 average is well below the US and Japanese scores) Almost all of the countries below Norway are the newer and/or poorer EU countries. More worryingly, Norway's performance on this synthetic indicator has deteriorated significantly over the years. Only Turkey has slipped faster, and most of the countries that were below Norway in 2006 have improved their performance (Figure 1.3). The synthetic indicator is built up from 25 individual indicators, which can be broadly divided into five categories, ranging from innovation drivers to IPR (Table 1.5). Interestingly, Norway scores high only on innovation drivers, for which the components include several of the framework conditions for innovation discussed below. Norway rates poorly in the other categories. It also performs poorly on converting innovation inputs into outputs.

The mediocre Norwegian results on the SII, as on the indicators discussed above (which are some of the components of the SII), seem puzzling at first glance, given Norway's undoubtedly good economic performance and the evidence from the total factor productivity data, which imply a high level of process innovation, especially in the services sector. One factor behind the low scores is that some (*e.g.* R&D spending) are calculated as a percentage of nominal GDP, and Norway's nominal GDP has increased strongly in recent years due to the rapidly growing value of petroleum exports. However, this can only be a short-term and partial explanation, as R&D and other innovation investments along with other components of the SII should be expected to increase along with income.

A closer examination of the sub-components of the SII index shows that:

- Within the "innovation drivers" category, and relative to the advanced EU countries, Norway scores:
  - High on population with tertiary education, broadband penetration rates, lifelong learning and youth educational attainment.
  - Low on new science and engineering graduates.
- In the "knowledge creation" category, Norway scores:
  - High on public R&D spending.
  - Low on business R&D, and share of medium-/high-technology R&D.

- In the "innovation and entrepreneurship" category, there are not enough reporting countries to make an accurate comparison, but Norway scores below Sweden and Finland in all sub-categories.
- In the "applications" category, relative to the EU15 average, Norway scores:
  - High in high-technology service employment.
  - Low in medium-technology and high-technology manufacturing employment, and very low in sales of new-to-market products and high-technology exports.
- In the "intellectual property" category, Norway scores:
  - Low in each sub-category (patent applications domestically or abroad, EU trademarks – even lower than the United States – and EU-registered industrial designs).

The picture that emerges is a consistent one: Norway has supportive framework conditions for innovation, especially as regards the skills level of the adult population, which helps to explain why its productivity performance is comparatively strong, especially in services. Financial support from the public sector is also a positive factor. But innovation activity in manufacturing has been weak. There is not much R&D spending, particularly in high-technology manufacturing, and very low sales of new-to-market products. The low scores on patent and trademark applications also support this hypothesis: manufacturing firms are more likely than service firms to seek patent protection, if only because goods are easier to define and describe precisely than services.

Insofar as innovation activity is weak in Norway, the weakness seems to be in the manufacturing sector. As noted above, low spending on business sector R&D can be "explained" by the country's industrial structure. When this is adjusted for, Norway's R&D spending looks comparatively robust in international comparisons. The fast growth of productivity, and especially of multifactor productivity, in the private services sector implies robust innovation activity in this sector. It is likely that standard indicators of innovation are less reliable for the services sector than for manufacturing.

Table 1.5. Components of the *European Innovation Scoreboard* summary indicator

Norway's rank in brackets

Inputs – (a) Innovation drivers:	
i)	Science and engineering graduates as a percentage of 20-29 age group **(low)**
ii)	Population with tertiary education as a percentage of 25-64 age group **(high)**
iii)	Broadband lines per 100 population **(high)**
iv)	Participation in life-long learning as a percentage of 25-64 age group **(high)**
v)	Percentage of 20-24 age group with at least upper secondary education **(high)**
Inputs – (b) Knowledge creation	
i)	Public R&D spending as a percentage of GDP **(medium-high)**
ii)	Business R&D spending as a percentage of GDP **(low)**
iii)	Medium-high- and high-technology R&D as a percentage of manufacturing R&D spending **(low)**
iv)	Percentage of enterprises receiving public funding for innovation **(high)**
Inputs – (c) Innovation and entrepreneurship	
i)	In-house innovating SMEs as a percentage of all SMEs **(low)**
ii)	Innovating SMEs co-operating with other SMEs as a percentage of all SMEs **(medium)**
iii)	Innovation spending as a percentage of turnover **(low)**
iv)	Early-stage venture capital **(medium)**
v)	ICT spending as a percentage of GDP **(medium-low)**
vi)	SMEs using organisational innovation as a percentage of all SMEs **(low)**
Outputs – (a) Applications	
i)	Employment in high-tech services as a percentage of total workforce **(medium-high)**
ii)	High-tech exports as a percentage of total **(low, including adjusted for oil)**
iii)	Sales of new-to-market products as a percentage of turnover **(very low)**
iv)	Sales of new-to-firm products as a percentage of turnover **(low)**
v)	Employment in medium-high and high-technology manufacturing as a percentage of total workforce **(low)**
Outputs – (b) Intellectual property	
i)	EPO patents per million population **(low)**
ii)	USPTO patents per million population **(low)**
iii)	Triadic patent families per million population **(low)**
iv)	New EU trademarks per million population **(very low)**
v)	New EU industrial designs per million population **(very low)**

Source: European Innovation Scoreboard 2006: Comparative analysis of innovation performance, www.proinno-europe.eu/index.cfm?fuseaction=page.display&topicID=5&parentID=51.

The policy implication is that there is considerable leeway for raising the level of innovation in manufacturing, the sector that will be mostly responsible for financing imports once the oil and gas runs out. The framework conditions are largely in place (see section 1.2), and the combination of low levels of protection and a strong real exchange rate should be putting pressure on firms to innovate to survive and prosper. If they do and if conditions are conducive to growth, they will grow. Large firms typically spend more on innovation than smaller firms. Up to a point, this may be because large firms were once small firms that spent a lot on research, and grew bigger as a result. But another reason is that their range of activities and scale of production can justify financing a research department. Consideration should be given to examining obstacles to firms' growth. As in many OECD countries, SMEs receive favourable treatment in terms of public support and employment regulations. But the support should not be so generous that it discourages expansion.

It seems clear that innovation activity in private services is healthy. It is not clear how much is imitation of best practice abroad (or at home) and how much is based on new ideas developed in Norway. But this is of secondary importance. What matters is that the sector is making active efforts to reduce costs and increase sales by adopting new techniques. As in other countries, this sector is expanding rapidly in terms of job creation – despite dynamic productivity gains – and value-added. It is important for the system of loans, grants and fiscal incentives for innovation to give equal treatment to this sector. This may necessitate changes in how requests for aid in this sector are vetted, given that it is often difficult to be precise about the expected gains in type or quality of service.

1.3. Framework conditions for innovation

Innovation is an issue of public interest for at least two reasons. First, it is widely accepted that successful innovation leads to better economic performance. Second, innovation activity is inherently risky, while the potential benefits to society from successful innovations may well be higher than the benefits accruing to the successful innovators. There is thus a case for policies that foster innovation, reduce the *ex ante* risks and costs for innovators, and ensure that the benefits are widespread.

This section looks at aspects of the Norwegian economy and policies that influence the level of innovation activity, positively or negatively. It focuses mainly on economic and socioeconomic features with an impact on innovation even where this is not their main *raison d'être*. Policies that specifically seek to boost innovation activity, such as tax breaks and subsidies, are less effective if the surrounding framework conditions are unsuitable. These framework conditions include a stable macroeconomy which encourages business to take a long view; a labour force with enough people with high-level training in science, engineering, mathematics and management; a labour market that allows employers to adjust the composition of their labour force as new products and processes are introduced; a product market that puts pressure on firms to innovate in order to survive and prosper; a financial market able to provide funding for risky projects and for firms new to the market; institutions to which firms can turn for advice and help when innovating and policies that encourage them to do so; and up-to-date transport and communication infrastructure that allows firms to acquire information easily and cheaply.

1.3.1. Macroeconomic stability

It may appear somewhat paradoxical that relatively stable macroeconomic conditions should favour innovation, the "creative destruction" that powers market economies and is the main source of their dynamism. Yet because innovation is inherently risky, entrepreneurs will be less ready to embark on new projects when there is additional risk due to unstable macroeconomic conditions – high and variable inflation, deep recession followed by overheating, unstable exchange rates, recurrent balance-of-payments crises, and volatile interest rates. In this respect, the Norwegian economy has had a reasonably good record. As Table 1.6 shows, during the past 35 years, per capita GDP growth has generally been somewhat faster and less variable than in a large sample of other OECD countries. Inflation rates have been variable, but less so than in trading partners, and Norwegian unemployment rates have been generally lower than elsewhere. The Norwegian mainland economy enjoyed a fourth consecutive year of exceptional growth in 2007, well above potential. At present, there is evidence that the economy has reached a cyclical peak, as rising cost inflation feeds into domestic price inflation (OECD, 2007d). Compared to earlier years, Norwegian firms increasingly report difficulties in recruiting highly skilled workers, including researchers.

Table 1.6. Macroeconomic indicators

	Per capita GDP growth								Inflation (CPI)			
	1970s		1980s		1990s		2000-05		1970s	1980s	1990s	2000-05
	Av.	Var.	Av.	Var.	Av.	Var.	Av.	Var.	Av.	Av.	Av.	Av.
Australia	1.5	1.1	1.5	2.6	2.4	1.5	3.1	0.8	10.4	8.1	2.2	3.0
Austria	3.5	2.2	2.1	1.4	2.1	1.2	1.8	1	6.3	3.5	2.3	2.0
Belgium	3.1	2.3	1.9	1.4	1.8	1.4	1.9	1.2	7.4	4.5	2.1	2.1
Canada	2.7	1.5	1.5	2.6	1.9	2.5	3	1.3	8	5.9	2	2.3
Denmark	1.9	2.4	2	2.1	2.2	1.5	1.7	1.3	9.8	5.9	2.1	2.0
Finland	3.3	2.7	2.6	1.6	1.6	4.3	2.9	1.3	11.1	6.6	1.9	1.2
France	2.7	1.5	2.1	1	1.6	1.4	1.9	1.1	9.6	6.3	1.7	1.9
Germany	2.7	1.9	2.2	1.6	1.7	1.5	1.3	1.2	5.1	2.6	2.4	1.6
Iceland	5.2	3.2	1.6	3.3	1.5	3	4.3	3		33.8	3.2	4.1
Ireland	3.3	2.1	3.3	3	6.3	3.3	5.9	1.9	12.7	9.1	2.3	3.5
Italy	3.3	2.8	2.4	1.3	1.5	1.2	1.7	1.4	13.8	9.6	3.8	2.4
Japan	3.2	2.6	3.4	1.7	1	1.6	1.7	1.2	9	2	0.8	-0.4
Korea	5.5	3.8	7.5	2	5.1	4.8	5.2	2.1	15	8.1	5.1	3.3
Netherlands	2.1	1.5	1.7	1.9	2.4	1.3	1.6	1.4	7.3	2.4	2.4	2.5
Norway	**4.2**	**0.7**	**2.1**	**2.1**	**3.1**	**1.2**	**3.3**	**0.9**	**8.4**	**7.6**	**2.3**	**1.8**
Portugal	3.6	5.2	3.1	3.3	2.7	2.1	1.2	1.7	19.1	17.1	4.9	3.2
Spain	2.6	2.9	2.6	2	2.5	1.7	3.5	0.8	15.3	9.3	3.9	3.2
Sweden	1.6	1.7	1.9	1.3	1.6	2.7	2.7	1.3	9.2	7.6	2.6	1.5
Switzerland	1.1	3	1.6	1.7	0.5	1.6	1.5	1.4	5	3.4	1.9	0.8
United Kingdom	1.8	2.7	2.5	1.9	2.2	1.7	2.7	0.7	13.7	6.6	3	2.4
United States	2.2	2.6	2.3	2.3	2.1	1.4	2.6	1.2	7.8	4.7	2.8	2.5
Average of the above countries	2.8	2.5	2.5	2	2.2	2.1	2.6	1.3	9.9	7.8	2.7	2.2

Note: Variability is measured as the standard deviation of the year-to-year growth rates over the period.

The brisk growth in production and exports of oil and gas in the past 10 to 15 years, and the more recent strong rises in their prices have created a macroeconomic shock that could easily have destabilised the economy. Many countries that begin to derive substantial rents from the exploitation of natural resources have suffered from a strong rise in the real exchange rate and a crowding out of the traditional tradable sector (the "Dutch disease"), gross inequalities in wealth distribution, low incentives to acquire human capital and corruption (the "resource curse"). To its great credit, Norway has avoided all of these. In particular, the creation in 1990 of the Petroleum Fund (later the Government Pension Fund – Global) as a fiscal policy tool to support long-term management of petroleum revenues was a key decision in this regard. The first net transfer to the Fund took place in 1996. Sizeable net transfers from around 2000 have so far prevented both an upward lurch in the real exchange rate (because hydrocarbon earnings are invested abroad) and fiscal irresponsibility *vis-à-vis* future generations (because export earnings are saved rather than spent). In the first few years of the 4% fiscal rule,[18] there was a significant overshoot in public spending out of the oil revenues, in part because the economy was operating below capacity and in part owing to adverse developments in international financial markets. Now, the use of petroleum revenues is substantially below the 4% rule, as economic activity is buoyant and the value of the fund is increasing rapidly. If oil prices remain very high, managing the oil wealth will become a greater challenge than in the past. Pressures to increase public spending on existing and new social programmes will become difficult to resist. Despite their famously healthy lifestyle, Norwegian workers have high and rising sickness and absence rates, which are proving difficult to combat, and pressures to finance early retirement remain strong. The experience of other countries in similar situations (*e.g.* the Netherlands in the 1970s and 1980s) shows that it is politically very difficult to scale back such programmes when they become harder to finance. On the other hand, high oil prices will undoubtedly also spur innovative exploration and recovery techniques in the offshore sector and partially offset the impact of increased hydrocarbon earnings on the R&D/GDP ratio.

18. The fiscal rule states that over time, the structural, non-oil central government budget deficit shall correspond to the expected real return on the Government Pension Fund – Global, estimated at 4%. However, the actual implementation of fiscal policy must take into account business cycle fluctuations around the suggested medium-term path. When capacity utilisation in the economy is high, this calls for fiscal policy restraint relative to the medium-term rule, whereas in a cyclical downturn somewhat higher spending of petroleum revenues may be justified.

1.3.2. Labour force issues – education

At least in the past, the Norwegian educational system has proved very effective in forming large numbers of well-educated adults. Full-time education is free until the upper secondary level and heavily subsidised at the tertiary level. Out of a population of a little over 4.5 million, nearly one million are in education, and over 200 000 are in tertiary education. Adults born before 1978 who have received education only up to the lower secondary level are entitled to upper secondary level education free of charge. More than 75% of the working age population has more than the compulsory level (currently up to age 16 in normal circumstances). As for other Nordic countries, the proportion of the population with tertiary education is at the high end of the OECD range for all age groups, and both men and women are likely to have spent over 13 years in formal education, with little gender difference (Table 1.7). According to the OECD's *Learning a Living* (OECD, 2005c), Norwegian adults have comparatively high levels of literacy, numeracy and problem-solving skills.

Table 1.7. Average years spent in formal education, 2004

	Total	Males		Females	
		25-34	55-64	25-34	55-64
Czech Republic	12.5	12.6	12.5	12.8	11.9
Denmark	13.4	13.6	13.6	13.6	13.0
Finland	11.2	12.5	8.5	13.5	8.5
France	11.6	12.8	10.3	13.1	9.6
Germany	13.4	13.6	13.7	13.5	12.5
Greece	10.9	11.9	9.4	12.6	8.2
Iceland	10.5	10.1	9.0	12.6	9.7
Ireland	13.0	14.0	11.2	14.5	11.4
Italy	10.1	11.2	8.7	11.7	7.6
Japan*	12.4	13.3	11.2	13.2	10.5
Korea	12.0	13.7	10.2	13.6	10.5
Netherlands	11.2	12.0	10.6	12.5	9.8
Norway	**13.9**	**14.2**	**13.4**	**14.7**	**13.1**
Poland	11.8	12.2	11.0	12.9	10.7
Portugal	8.5	9.3	7.3	10.3	7.2
Slovak Republic	12.5	12.8	12.1	13.0	11.3
Spain	10.6	11.9	8.9	12.5	8.0
Sweden	12.6	13.1	11.3	13.6	11.8
United Kingdom	12.6	13.0	12.4	12.9	12.0
United States	13.3	13.1	13.2	13.4	13.1
OECD average	**11.9**	12.5	11.0	12.8	10.3

*Japan: 2003.

Source: OECD (2007c), Education at a Glance 2006.

More recent educational results show an unwelcome trend among the lower age groups, however. Although Norway spends a higher than average share of GDP on education, evidence from the OECD's PISA programme, including the findings published late in 2007 (see Box 1.4) as well as other sources, indicates that Norwegian schoolchildren today lag behind those in other countries as regards mathematical ability; if anything, the situation has worsened over the past decade (Table 1.8). Dropout rates in lower and upper secondary education have risen, as has the proportion of students failing to complete higher education courses. Similarly, the percentage of graduates from upper secondary education with a mathematics, science and technology (MST) orientation was barely above 20% in 2003, having fallen from about 28% nine years earlier. Most comparable countries have figures above 30% (Norwegian Ministry of Education and Research, 2006).

Box 1.4. Results from the OECD PISA assessments

As part of its ongoing work in the field of education, the OECD has, since 2000, conducted triennial assessments of the educational performance of students in the 15-16 year-old age group, those typically approaching the end of compulsory education. The "Programme for International Student Assessment" (PISA) focuses only partly on how much knowledge has been acquired by students: the main focus is on their ability to use the knowledge they have acquired (reading, mathematics and scientific literacy). The first PISA assessment focused on reading literacy, the second on math competencies (numeracy), and the most recent, conducted in 2006, focused on scientific competencies. A representative sample of several hundred thousand students from all OECD member countries, and from an increasing number of non-OECD economies, participates in the assessments. The sample contains students from all types of schools and localities. Participants are given a series of written tests (the same for all students), with multiple choice questions, questions which involve choosing the correct response and giving reasons, and questions requiring a written response. The tests are designed by educational experts from countries participating in the assessments, and the results are scored uniformly across schools and countries. No attempt is made at this stage to make allowance for differences in the number of years of compulsory schooling, spending on education, type of school, class size, or socio-economic backgrounds. The PISA publications present analyses that estimate to what extent such factors influence the results. In general, the differences in scores within and between countries are significantly greater than can be explained by such factors, implying that there are substantial differences among individual schools within countries, and among educational systems across countries, in their effectiveness in preparing students for the challenges they will face as young adults in a globalising world.

Successive PISA assessments, based on the raw score data, have typically put Norway at or somewhat under the middle of performance among OECD countries (the assessment reports are rich and detailed, and the foregoing summarises only a small proportion of the more important findings). The PISA 2000 assessment of reading literacy placed Norway above France, Denmark, Italy, Germany and the United States, but below other English-speaking countries, Finland, Sweden, Iceland, Korea and Japan. Broadly similar results were obtained for the assessment of mathematical and scientific literacy. In all three cases, Norwegian student performance in the 2000 assessment was not significantly different from the OECD average.

Box 1.4. Results from the OECD PISA assessments *(continued)*

The 2003 PISA assessment of mathematical ability – especially the ability to use mathematical knowledge to solve real-world problems – showed no significant change in Norway's relative position since 2000 in areas that were tested in both assessments. Overall, though, in this assessment, Norwegian student performance was slightly but significantly lower than the average. It ranked 21st out of 30 OECD countries: several of the countries that scored lower than Norway are classed as developing countries. Analysed in slightly more detail, above-average performance in questions of a statistical nature did not offset below-average performance in questions relating to shapes, changes and quantities.

The latest PISA assessment, details of which were released late in 2007, focuses on students' ability to use scientific knowledge and apply the scientific method to answer a variety of questions related to various aspects of the Earth and space, living systems and physical systems. The countries covered comprise about 85% of world GDP, and included major non-OECD countries such as China, Russia, Indonesia and Brazil, as well as several smaller ones.

The results for Norway were at best mediocre. Norwegian students performed significantly below the OECD average, being ahead of only Luxembourg, Italy, Greece, Portugal, Turkey and Mexico in the OECD area (and below several eastern European countries and China). Except for Finland, Norway's Nordic neighbours also had only average scores, or even slightly below (Iceland). By contrast, Finland was rated the top-scoring country by a comfortable margin, with only around 5% of students with scores in the two lowest levels of ability (the figure for Norway was around 20%), and about the same proportion in the very highest ability level, number 6 (about 1% for Norway). Although there is a rough relationship between the PISA scores for 15/16 year-old students and the proportion of adult scientific researchers in the population, a peculiarity of the Nordic countries – including Finland – is that they have considerably more researchers than would be expected, given the PISA scores for students. This may imply that these countries are unusually successful in eventually training large numbers of researchers despite mediocre school performance (except for Finland), or that the quality of scientific teaching has declined in these countries (except for Finland).

When the raw scores are adjusted for such things as educational spending and per capita GDP, the Norwegian results look even worse. Although there is no reason to believe that students in rich countries are inherently better at science than those in poorer countries, rich-country science students will have better access, other things being equal, to laboratory equipment, modern textbooks, information services and other teaching aids. Furthermore, spending on education relative to GDP is not identical across countries. The PISA analysis shows, for example, that adjusting the raw scores for per capita GDP would put Norway's performance below that of Greece, and only slightly above Turkey. Similarly, given Norway's per capita spending on education, its PISA score "should be" about 20 points higher, putting it ahead of Sweden and Denmark.

.../...

Box 1.4. Results from the OECD PISA assessments *(continued)*

The PISA scientific assessment examined separately three aspects of the ability to apply scientific knowledge: identifying scientific issues; explaining phenomena scientifically; and using scientific evidence. Relative to its (mediocre) overall score, Norwegian students scored comparatively well on the first two (though not better than the OECD average), and particularly poorly on the third, with about one-third of students in the two lowest levels (among OECD countries, only Italy, Turkey and Mexico had worse performances). By contrast, the countries with the highest overall scores all did particularly well on the use of scientific evidence. Of the large number of other factors and influences that were discussed in the PISA report (about 800 pages of analysis and data), Norway stands out in two respects: for the countries sampled, Norwegian school principals reported the least competition for students from other schools in the same locality; and virtually all Norwegian 15-16 year-olds are following at least some science courses. The OECD average is under 90%, hence the Norwegian sample presumably includes students of low scientific ability. These two factors could go a little way in explaining the less than satisfactory performance of Norwegian students.

A possible reason for these disappointing results is the lack of appropriately qualified schoolteachers in mathematics and science, in part because, in Norway as elsewhere, technically qualified persons can earn higher salaries outside the teaching profession, but also because it is not obligatory to hold a university degree or equivalent in mathematics or science to teach up to the lower secondary level.[19] As a result, a significant number of students entering higher education for non-MST courses are unable to perform simple calculations. The publication "A Joint Promotion of Mathematics, Science and Technology" (Norwegian Ministry of Education and Research, 2006). *i.e.* well before the latest PISA results became available, highlights the low level of interest in following science-oriented studies not only among Norwegian schoolchildren, but also trainee teachers, and the resulting dearth of adequately trained school teachers in the relevant disciplines. Only 1-3% of primary school teachers were participating in continuous training in mathematics and science. Measures have been taken to address these issues. Extra financing will be available to train future secondary school teachers in mathematics, science and technology,[20] and the number of applications to MST teacher training programmes is reported to be increasing. This is a welcome development, as more than half of all current MST staff with higher education qualifications are over 50 years old. Lower down the age

19. Compulsory mathematics courses were introduced into the general teacher training programme in 2003.

20. Norwegian science studies present a mixed picture, with some strong groups and some weaker ones. However, the trend has been favourable over the past decade, with comparatively rapid developments in ICT and growing numbers of publications and citations in reputable journals (see section 3.3.2.2 in Chapter 3).

scale, the intention is to increase funding and encourage the teaching of MST at the primary and secondary school level. The reform on "culture for learning" and "knowledge promotion" (*Kunnskapsløftet*) aims to devote more teaching time to MST, promote gender equality and raise the proportion of upper secondary students following MST courses. The criteria for entry to universities (including university colleges) for older students (over 25 years) who have not completed upper secondary education will be amended to permit them to attend if their non-formal qualifications are deemed adequate.

Table 1.8. Achievements in secondary education

	(I)	(II)	(III)	(IV)	(V)
		High benchmark		Intermediate benchmark	
		1995	2003	1995	2003
Czech Republic	84				
Denmark	85				
Finland	93				
France	83				
Germany	76				
Greece	62				
Iceland[1]	85				
Ireland	83				
Italy[1]	69	26[3]	23	59[3]	59
Japan	86	54	53	85	86
Korea	91	50	57	81	88
Netherlands[1]	88	48	43	82	85
Norway	**80**	**32**	**21**	**72**	**63**
Poland	78				
Portugal	70				
Slovak Republic	81	42	34	77	72
Spain	76		20		58
Sweden	83	52	38	83	75
United Kingdom[2]		30	32	61	70
United States	74	38	41	68	75

Notes: Column I: percentage of 15-year old students achieving levels 2-5 on maths scale; Columns II-V: percentage of 4th and 8th grade students reaching TIMM benchmarks of science achievements.

1. 2002. 2. Scotland. 3. 1999.

Source: OECD (2003), *Learning for Tomorrow's World. Trends in International Mathematics and Science Study.*

Policy efforts like these are all the more urgent because the proportion of students opting for relatively demanding science and technology courses at the bachelor and higher levels is not high and falling (Table 1.9). Although in Norway as in other OECD countries the number of students in higher education has risen in recent decades, the proportion studying mathematics, science and technology has fallen (Figure 1.4) even though there was an increase in applications to engineering studies from 2006 to 2007. Absolute numbers remained stable until 2004, followed by a drop, although it is relatively easy to enter such courses (2006 results pointed to a recovery in numbers of students entering the technological fields, but the numbers applying for maths and science continued to decline). A similar phenomenon is observed in other advanced OECD countries, including Norway's Nordic neighbours, but this offers little consolation. Students acquiring advanced technical skills are the research personnel of tomorrow. At present, Norway still has adequate numbers of such trained adults in the workforce,[21] a necessary condition for brisk innovation activity. If the supply dries up, other policies designed to encourage innovation will not succeed. Consideration should therefore be given to strengthening existing and planned incentives for teachers and students to acquire and practice MST skills. The PISA studies show that schools can learn from best practice in similar schools in the same country, and that national education systems can profit best practice systems in other, similar, countries.

Table 1.9. Enrolments and graduation in higher education

Academic year	1994/95	2000/01	2001/02	2002/03	2003/04	2004/05
Numbers in tertiary education in Norway and abroad						
Total	181 736					224 158
Business and administration	28 251					41 284
Graduates with degrees of duration 4 or fewer years						
Total	19 835	23 975	22 851	23 161	23 799	23 475
Scientific and technical	3 024	3 547	3 239	3 396	3 289	2 921
Graduates with degrees of duration longer than 4 years						
Total	6 323	7 205	6 732	6 726	7 605	7 848
Scientific and technical	2 442	2 245	1 856	1 830	2 321	2 005
Business and administration		503	644	542	1 172	1 145
Doctoral degrees						
Total	605	768	740	714	756	838
Science and technology	290	332	320	316	298	385
Business and administration		16	18	15	22	17

Source: Statistics Norway.

21. Business-sector researchers as a proportion of all industrial employment is not the highest among OECD countries (the United States, Japan, Sweden and Finland are ahead of Norway), but it is well above average. At the time of writing, the generalised tightening of the labour market was leading to difficulties in recruiting and retaining skilled researchers.

Figure 1.4. Science and engineering degrees, 2004[1]

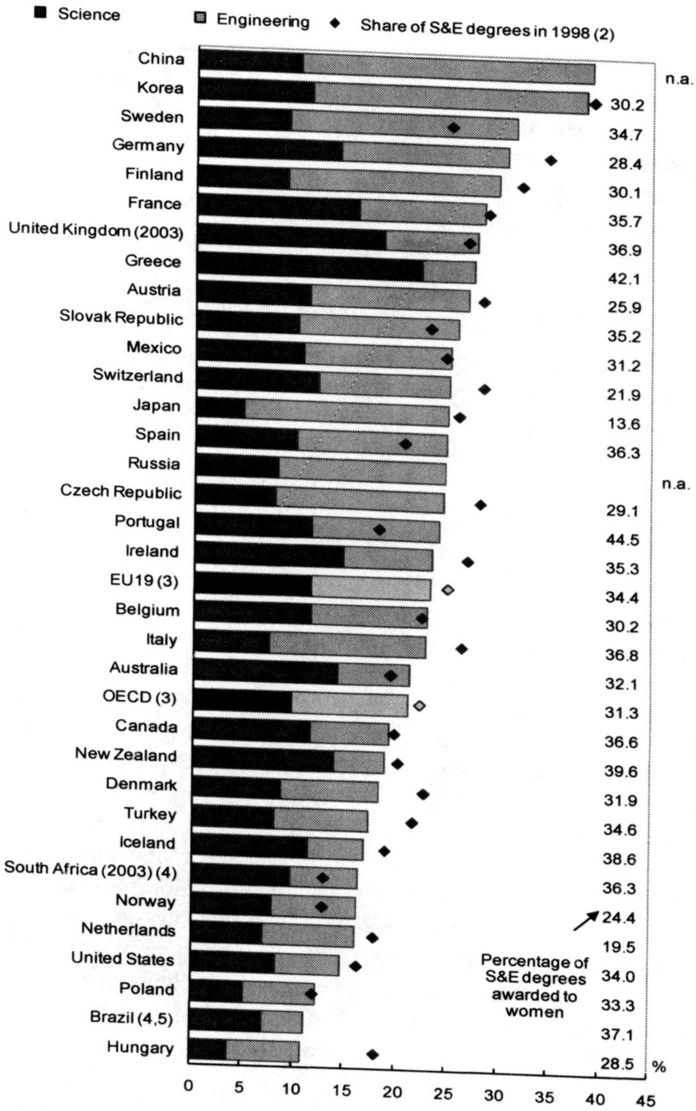

1. As a percentage of total new degrees.

2. 1999 instead of 1998 for the Slovak Republic and Denmark; 2000 for Portugal and Belgium. These four countries as well as Greece and Luxembourg are excluded from the calculations of EU19 and OECD in 1998.

3. Excludes Luxembourg. 2003 data for the United Kingdom.

4. ISCED 5B programmes are included with ISCED 5A/6.

5. Share of S&E degrees awarded to women is for 2003.

Source: OECD Science, Technology and Industry Scoreboard 2007.

In addition to intensifying efforts to raise the number and quality of MST students in full-time education, the authorities are making efforts to step up the quality of Norwegian research at the university and equivalent level. To this end, Norwegian Centres of Excellence have been created in some universities. They are composed of research teams deemed to be of high quality and receive increased funding. More financial support is being given to Norwegian researchers studying abroad, and academic researchers are encouraged to profit financially from their research via the FORNY programme and the establishment of technology transfer offices (TTOs) in universities.

1.3.3. Labour force issues: training, mobility and flexibility

The flexibility and mobility of a country's labour force can have an important impact on the intensity and direction of in-house training, research and innovation, but the issue is complex. By definition, innovation involves new processes and/or new products, and unless the innovation takes place in a new firm, employees may need to acquire new skills, move to different positions within the enterprise, and possibly be laid off if their skills are made redundant or if the innovation is labour-saving. In countries where laying off employees can be expensive (as in several continental European countries), there will be a combination of incentives to minimise the size of the labour force, by adopting capital-intensive techniques and outsourcing, plus incentives to maintain training programmes for employees, so that their skills remain useful as technology and processes change. In such environments, employers also have incentives to concentrate training in areas that are specific to the firm, so that employees will be less tempted to leave when their training is completed and work elsewhere for higher salaries. These firms are more likely to engage in original R&D. In countries where it is comparatively easy to alter the size of the labour force and move workers around the firm, spells of unemployment are generally comparatively short, as employers are more willing to hire. They have less incentive (at a given unit labour cost level) to invest in capital-intensive techniques and less incentive to pay for training their employees in techniques specific to the firm or indeed in anything else. In such cases, training may need to be subsidised or organised collectively at the industry level.

In Norway, based on the standard measures developed by the OECD, the labour market appears to be moderately flexible. Dismissal for fault is comparatively expensive and time-consuming, but no-fault dismissal of individuals is relatively fast and cheap. Unlike many other European countries, Norway has no significant "dual labour market" problem, with highly protected employees on permanent contracts while those with temporary or short-term contracts have little or no protection. Low unemployment means mployees can be reasonably confident that if they are laid off, they will kly find another job. Long-term unemployment accounts for about one-rter of total unemployment, compared to 60% in the EU. In the current rwegian environment of low unemployment and positive employment owth, employees may be less averse to accepting changes in their working conditions than in other countries, making it easier to carry out process and product innovations. Pressures to protect workers in the inevitable economic downturns should be resisted. Continental European countries that followed this route in the late 1970s and early 1980s found themselves with high rates of structural unemployment and high NAIRUs[22], which ensured that recoveries were short-lived.

Statistics Norway's *Urban and Regional Labour Markets: Mobility in Norway* gives further evidence of mobility trends in Norway. Over half a million workers change jobs each year on average, out of a total labour force of some 2.5 million. Most changes occur within sectors, with workers quitting one job to take another. Job changes resulting from entry into or out of unemployment, or resulting from firm closures or start-ups are much less important. As might be expected, persons with high levels of education are more likely to pass quickly from education or unemployment to a job than people with low qualifications. Job churning is more intense in the Oslo region than elsewhere at all levels of education.

Another feature of the Norwegian labour market is the very high employment rate: about 75% of the working-age population is in employment, 10 percentage points above the EU average. A striking 70% of older workers (55-64) are employed, against an EU average of 45%. In addition, youth unemployment is only around 12% compared with an EU average of 16% (and in Norway most of these are in fact part-time students). One consequence of the high employment rate is, of course, a high level of per capita output, which is substantially modified owing to the importance of part-time employment, especially among women: annual hours worked are well below the EU and OECD average. Another consequence is that compared to other OECD countries, a relatively large share of people with limited skills and accomplishments are employed. Since wages are high in real terms and their distribution is flat, employing workers with marginal skills is feasible only if their productivity is high or if their employment is subsidised. This could encourage employers in the private sector to find innovative ways of raising the productivity of the low-skilled.

22. The NAIRU is the non-accelerating inflation rate of unemployment. When it is high, even mild upturns will lead to higher inflation even though registered unemployment is high.

1.3.4. Competition in the product market

An economy needs sufficient competition for vibrant innovation although the interaction between innovation and competition is quite complex (see Box 1.5).

Box 1.5. Innovation and competition

There is a complex relationship between innovation activity, firm structure and the intensity of competition among firms, which involves firm dynamics. There are two opposing factors: competition potentially reduces profits, which are needed to finance innovation. The ideal firm of static economic theory (which has never existed outside the textbooks) has no incentive to invest in R&D because this is costly for itself, but competitors by definition will get the results free of charge. Nothing changes in this world, except the weather and the harvest.

By contrast, Marx and, in the 20th century, Schumpeter emphasised the dynamic aspects of capitalist enterprises, the process of "creative destruction" in which firms continually seek market power by weakening their competitors via the introduction of new products or cheaper processes. In this world, resource allocation is always sub-optimal in the above textbook sense. But as long as the profits are ploughed back into further innovation, this is compensated by the power of innovation to create new resources and raise efficiency. Hence monopoly power can be tolerated, indeed encouraged, as long as the firms with market power use their deep pockets to finance innovation. On this view, the competition policy authorities should take action only against firms that are using their excess profits to enrich owners that settle for a quiet and lucrative life.

The possibilities for this kind of market structure depend on the industry. Even at world level, there may not be room for many airplane manufacturers or major pharmaceutical companies, whose continued existence depends mainly on successful innovation. But there are other industries with large firms in which the possibilities for innovation are less evident. At least in the past few decades, there have been few innovations in the primary extraction industries (excluding offshore petroleum), where giant firms are the norm, or in the manufacture of bricks and cement, where again large firms dominate. At the other end of the spectrum, the retail footwear and clothing sector has very large numbers of very small outlets, and a few giant chains with outlets in cities throughout the world. None has real market power, and they all compete by offering innovative products or services. Market power is neither a necessary nor a sufficient condition for innovation activity. Empirical evidence, as summarised by Ahn (2001, 2002), shows that:

- Competition has a lasting dynamic impact on firms' behaviour which is not captured by conventional short-term static analysis.
- In some high-technology sectors characterised by network effects and positive feedback, conventional measures of the intensity of competition, such as concentration ratios, may imply little competition, whereas in practice competition may be fierce.
- There is no clear-cut relationship between market concentration or firm size and innovation activity.

…/…

Box 1.5. Innovation and competition *(continued)*

- There is a robust relationship between product market competition and productivity growth (which is related to innovation activity).
- Greater intensity of competition (*e.g.* through regulatory reform, opening of markets to foreign suppliers) results in an increase in productivity growth and higher consumer welfare.
- Competition between existing firms is important, but competition from innovative new firms may be more important in securing productivity gains at the cutting edge of technology.
- The interaction of competition in product, labour and financial markets has an important influence on innovation and growth. In particular, narrow illiquid capital markets and inflexible labour markets hold back most types of innovation activity, except imitation of technologies already becoming obsolete in more advanced countries.

In practice, the type of product market competition also favours particular types of innovation activity. In the typology developed by Aghion and Howitt (2005), firms may find themselves in different competitive and technological environments. At one extreme, institutions and laws discourage competition in all markets (*e.g.* by giving producers exclusive rights to set prices and choose outlets), and there is little incentive to invest in costly and uncertain original innovation activity. A low-cost alternative is leasing and licensing processes and the manufacture of obsolescent products from firms in more advanced countries. The further a country is behind the technological frontier, the greater the incentive to buy innovations elsewhere and market them at home. Producers' profits can be good and consumers have little choice because tacit or explicit collaboration among firms to divide up the market is common. Provided that social, political and economic institutions are conducive to growth, productivity gains can be large as the economy progresses towards the productivity frontier and GDP growth is rapid. But once the country or sector approaches the frontier, the cost of this approach rises, and catch-up slows or halts. Some European countries had this kind of market structure for a few decades after World War II and some developing countries have it today.

At an intermediate, somewhat more advanced, stage, it becomes clearer that more intense competition helps raise efficiency and encourages innovation. Competition authorities are given the power and the means to break up cartels and to punish firms that engage in collusion. They can also prevent mergers that might result in excessive market power and forbid governments to step in to aid failing enterprises. This competitive environment characterises EU countries today, although the speed and conviction with which anti-competitive activities are prosecuted by the authorities varies across time and across countries. This environment creates a dilemma for established firms at some distance behind the most advanced technology in their field. If they innovate to catch up with the leader, competition between them will reduce the current leader's profits. They also face potential competition from new entrants with even more advanced technology. If they innovate but there are such new entrants, their innovation will earn them no extra profits. While they will therefore have little incentive to innovate, they can expect to earn normal profits on their existing products. However, as they gradually sink down the technology ladder, their market will shrink, and so will their profits.

…/…

Box 1.5. Innovation and competition *(continued)*

Firms that are close to the frontier have more incentive to innovate, especially if there is a real threat of entry by new, technologically sophisticated firms. If they do not innovate and these entrants materialise, their profits will dwindle. They can escape this to some extent by innovating and earning above-normal profits for a while, but the process will repeat itself. Hence when it is relatively easy for new, technologically advanced firms to enter the market, innovation activity will be brisk, the technological level will be high and the productivity level will also be high. This is the case in the ICT sector. Firms at the leading edge have to "run to stay in place". The threat created by potential new entrants extends beyond the ICT sector. When it is easy and cheap for new firms to enter a market, market leaders in that market will have an incentive to innovate but firms well below the frontier will not. There is thus likely to be a wide spectrum of productivity levels and profit rates within the industry, but even the laggards can still make a profit on their obsolescent processes and products. This helps explain the empirical finding that industries are usually characterised by wide variations in firms' productivity (*e.g.* Oulton, 1998).

In sum, in countries and sectors where firms are mostly below the technology frontier, lack of competitive pressure will result in lack of original innovation. Firms can make good profits by acquiring technologies that are already below best practice. When at least some firms are close to, or at, the technology frontier, the spur to innovation will come mainly from new firms with cutting-edge technologies (but which lack marketing experience). In that situation, incumbents must innovate or die. If there is little threat of entry, the spur to innovate is weakened, although competition among incumbents close to the frontier will have some effect.

It is not easy to judge the importance of the various forces in Norway described in Box 1.5. The Norwegian Competition Act of 2004 contains provisions very similar to those of the European Economic Area (EEA) and is largely harmonised with EU competition rules. The Norwegian Competition Authority (NCA) is under the authority of the Ministry of Government Administration and Reform, which is the appeals body, but has no power to dictate outcomes of particular cases. Fines levied by the NCA can be challenged only in the courts. As in the EU, the potentially positive dynamic impacts of innovation are taken into consideration when corporate behaviour would otherwise be judged illegal. Symmetrically, behaviour that is not otherwise illegal can be ruled to be so if it is held to restrict innovation.

Conventional measures of the intensity of competition among firms in the product market give average results for the Norwegian non-manufacturing sector[23] (Table 1.10). Norway was rated at about the same level as Germany, Finland and Denmark, well ahead of France and Italy, but behind Sweden and especially the United States and United Kingdom. Norway

23. The usual assumption is that the intensity of competition is high in manufacturing because of the virtual absence of tariff and non-tariff barriers.

performed relatively poorly in a recent measure of the strictness of competition law and policy (Høj *et al.*, 2007) but this may change owing to recent legislation. In particular, the current competition law mirrors that of the EU, and this may affect the rankings. The fact remains, however, that Norway stands out among OECD countries in terms of public sector (municipal as well as central) government involvement in industry through financial participation or outright ownership (OECD, 2007d). Such involvement is conventionally held to dampen competition among firms in the private sector. The Norwegian authorities argue that this is not the case in Norway, as firms that are wholly or partly state-owned do not have privileged access to capital or markets, are subject to the same rules (including competition policy rules) as other firms, and are not saved from bankruptcy if they become financially non-viable. Barriers to inward FDI are not high (except in some sensitive sectors) and FDI inflows relative to GDP are average (Figure 1.5), suggesting that foreigners are not overly discouraged by the degree of state ownership or other factors.

Table 1.10. Summary indicator of product-market competition, 2003

Country	
Australia	0.88
Denmark	1.12
Finland	1.32
France	1.70
Germany	1.43
Iceland	0.97
Ireland	1.12
Italy	1.87
Japan	1.28
Norway	**1.48**
Sweden	1.23
Switzerland	1.68
United Kingdom	0.92
United States	1.03

Note: Higher numbers indicate higher barriers to competition.

Source: OECD Product Market Competition database.

Figure 1.5. Foreign direct investment inflows

Percentage of 2003-05 GDP

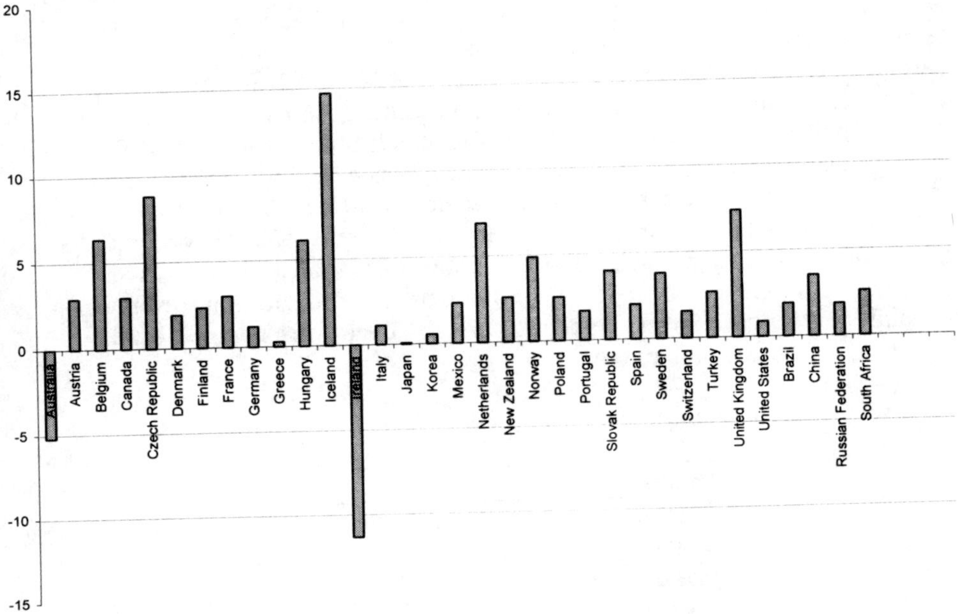

Note: Profits on FDI, whether or not remitted, are counted as negative inflows. The GDP denominator for Norway is the total economy, not mainland Norway, as the offshore sector also attracts inward FDI flows.

A sub-set of the same general set of indicators shows that barriers to entry for new firms are on a par with those in the best-performing countries (Table 1.11) and, like the general indicators described above, have improved over time. Objectively, and although there may be lingering qualms about the effect on competition of the pervasive presence of the public sector in industry, the conditions for competition between existing firms and from potential new firms, seem to be conducive to innovation activity.

Table 1.11. Barriers to trade, foreign investment and start-ups

Country	Barriers to trade and foreign investment	Barriers to start-ups
Australia	0.89	1.0
Denmark	0.84	0.5
Finland	0.64	1.3
France	0.95	1.9
Germany	0.64	1.6
Iceland	0.30	1.4
Ireland	0.54	0.5
Italy	1.15	2.4
Japan	0.93	1.9
Norway	**0.79**	1.0
Sweden	0.79	1.2
Switzerland	1.01	1.7
United Kingdom	0.36	0.7
United States	0.73	1.0

Note: Higher numbers indicate higher barriers.

Source: OECD Product Market Competition database (2003 data).

Nevertheless, although barriers to entry for new firms are comparatively low, the share of self-employed persons of working age seems to be small given the ease of entry.[24] This mostly reflects the low share of necessity-driven entrepreneurship (see section 2.3.1.1). Another factor might be the comparatively flat income distribution in Norway. One reason for becoming an entrepreneur, with its attendant risks, responsibilities and uncertainties, and apart from the satisfaction of being one's own boss, is the possibility of becoming much wealthier than one could be as an employee. This may be less attractive in Norway, where the tax system makes it less possible, and where ostentatious displays of private wealth are not universally regarded as admirable.

24. The rate of self employment is 7.5% in Norway, compared with an EU average of close to 16%. Rates in Denmark, Sweden and Finland are also below the EU average. However, the figure for Norway is identical to that for the United States.

1.3.5. Financial markets and innovation

The structure and depth of financial markets is important to the innovation process. Innovation in the financial sector – the spread of ATMs, Internet banking and electronic fund transfer facilities, real-time capital market transactions and portfolio management – has made it far quicker and cheaper for individuals and institutions to manage their finances, and has certainly improved the allocation of investment finance, an important factor in boosting long-term growth rates. Most CEOs in Norway's financial sector regard innovation as "very important".[25] The sector also plays an important role in facilitating innovation in other sectors. Innovation is inherently risky and may be lengthy and costly. Even if successful, a new product or process may come on stream during an economic downturn. Large firms with some market power can of course finance their in-house research teams out of retained profits, in the reasonable expectation that they will at least break even over the longer term, but even they may prefer to borrow at comparatively low interest rates and return profits to shareholders, possibly boosting their share price as a result and reducing future capital costs.

Moving down the size scale, conventional commercial bank finance is likely to be costly or unavailable even for modest established firms that want to move into unfamiliar areas. Equity finance is more appropriate: the higher price is one the firm may be willing to pay. However, small firms, especially new ones, may be unable to finance (risky) research out of their own resources or to tap the traditional public equity markets. Start-up firms may have little managerial experience, little collateral and merely a promising idea for which there is as yet no market. What they need is venture capital which can spread risk over many projects and supply financial and managerial expertise.

The Norwegian equity market is relatively underdeveloped. Private equity investment is equivalent to about 15% of GDP compared with a European average of around 25% (and over 80% in Sweden). A very high proportion of equity is held by the corporate sector; private holders account for less than 5%.[26] The low overall level of private equity capital in Norway is in large part a reflection of the level of state ownership which is partly compensated by the state financial contribution to innovation activity via the funds administered by the RCN and Innovation Norway. Like suppliers of equity finance, the state entities evaluate projects and help finance those

25. Oral communication from DNBNOR.

26. Banks, insurance companies and pension funds are generally less able to invest in private equity than is usual in other OECD countries. Together, they account for around 20% of private equity capital compared with an average of 60% in the EU.

they judge to have a good chance to succeed (including those that seek to benefit from the tax credit scheme, Skattefunn). Since they are dealing with public money, state suppliers of finance for innovation activity may be more conservative when lending to the private sector than suppliers of private equity in other countries would be. Innovation projects for which it is easier to make cost estimates and which have a definable physical end-product, projects typical of the manufacturing sector, might be more likely to receive finance than those in the services sector. Also, because there are several ministerial suppliers of finance to the RCN and Innovation Norway, problems of co-ordination may arise, with too much finance going to "pet projects" or to more than one project in the same field.

The Norwegian public sector also actively helps finance the smallest projects, *e.g.* in start-up firms. On behalf of the Ministry of Trade and Industry, Innovation Norway administers the Nationwide Seed Capital Scheme which can lend up to NOK 667 million to investment funds located in four university cities, and the Regional Seed Capital Scheme which can lend up to NOK 700 million to investment funds oriented towards start-ups in assisted areas. In addition, the government owns a fund-of-funds investment company, Argentum, established in 2001 with a capital base of NOK 2.45 billion which subsequently received a further NOK 200 million. Its objective is to facilitate access to foreign venture capital and to encourage the development of the Norwegian equity market. All of its investments require majority private ownership, and state participation is hands-off. It seeks to enhance competitiveness in the Norwegian business sector and to achieve a high yield on the capital invested. It is also tasked with strengthening competent, long-term ownership in the Norwegian business sector and the creation of networks involving owners, fund managers and R&D centres. Although the absolute size of the capital available from these three entities is large, the total is only about 0.1% of GDP.

Norway's private venture capital market presents a mixed picture. It is dynamic, it has ample funds at its disposal, its asset base is considerably larger than that available from public sources, but it is nevertheless quite small (see Chapter 2, section 2.5).

1.3.6. Institutional knowledge sharing

Innovation activity flourishes best when information about new developments and techniques is readily available. In the academic world this is ensured by the pressure to publish new findings, the vast array of relevant journals and websites, and the conferences and seminars at which new ideas are presented, discussed and evaluated by experts in the field. Information sharing in the private sector is a more complex issue. Firms usually want to keep commercially sensitive information concealed from their competitors until it is protected by law or until it has become a new product or process with a substantial lead time over rivals. Depending on how strictly competition policy is applied, it might even be a criminal offence in some countries to discuss market-sensitive information with competitors.

Norway is a comparatively small country with a tradition of co-operation. Although it is difficult to adduce quantitative proof, it is at least plausible to argue that there is a significant amount of informal information sharing of a technological character between firms in the private sector. As noted above, there is also a considerable amount of job churning; employees often leave one firm to join another in the same sector. Given that only the petroleum sector is very large, information about new developments is likely to circulate via this informal route. In addition, the state is an active supplier of information and advice, as well as loans and grants, on innovation activity. The Research Council of Norway is a meeting place for researchers in the private and public sectors as well as a distributor of funds for R&D. In certain areas, grants from the RCN may be conditional on the recipients co-operating with national or foreign institutes and researchers. The recently created Centres for Research-based Innovation encourage information sharing and contacts between R&D-intensive firms and major research groups. The RCN also co-operates with Innovation Norway and SIVA in providing innovation services to the private sector.

Quite apart from its role as a supplier of grants and loans, including seed capital for new companies, Innovation Norway also plays a major role in encouraging innovation activity by other means. For example, it encourages Norwegian SMEs to collaborate in the development of an innovative product, service or process that is new to their market. This takes place under an industrial R&D contract (IFU) and normally involves at least two SMEs with a supplier-customer relationship. Financial support may be easier to obtain if the Norwegian firm collaborates with a foreign enterprise in developing a new product or process.

SIVA, the Industrial Development Corporation of Norway, provides practical information and infrastructure services for innovation activity. It has part ownership of science and research parks and can advise firms as to where to find new information. Part of its mandate involves the creation of national and international networks of R&D centres.

In sum, innovation activities in the private sector in Norway are not hindered by lack of access to relevant information. Formal and informal networks exist for this purpose, and the formal ones receive substantial public support. In addition, the generally high level of education among Norwegian adults, and the near-universal access to Internet services means that relevant information circulate and its relevance is appreciated.

1.3.7. Public procurement

The scale of public procurement is very significant. In 2006, public purchases of goods and services amounted to NOK 315 billion. Purchases by general government units amounted to NOK 232 billion and the remaining funding is by government enterprises, including in the oil sector. Norway's public procurement system is decentralised;[27] the Ministry of Government Administration and Reform is in charge of procurement law. Decentralisation is expected to increase the scope of suppliers and induce fiercer competition. At the same time it is difficult to staff all parts of the system with the "intelligent customers" needed to operate it at a high level of efficiency. There are efforts to provide specialised education and training as well as examples of co-operation among independent procurement agencies.

The procurement directives are in line with EC directives and do not impede innovative procurement. However, there is strong emphasis on efficiency and transparency, rather than on explicit measures to stimulate research and innovation through procurement. Procurers are assessed against a number of criteria, but the stimulation of research and innovation is not among them.

There is reviving international interest in demand-side innovation policies based on procurement, not least in the European Union. A European expert panel report (European Commission, 2005) on innovative procurement argues that the gains from innovative procurement can be realised under the new European directives for public procurement. Opportunities exist within:

- Negotiated procedures and competitive dialogues, which can be used optionally to structure the procurement process in certain situations and to facilitate the critical element of dialogue between customer and supplier.
- Technical dialogues in the preparation phase before tenders are sought.
- The equal footing now given to technical specifications made in terms of functional or performance-based requirements and to references made to standards.
- Options to submit variants.
- Conditions that allow transfer of intellectual property to the supplier.

27. There exists no central body for public procurement in Norway apart from public procurement activities for hospitals. Procurement Services for Health Enterprises is the company responsible for co-ordinating public procurement on behalf of Norwegian health enterprises.

The panel concludes that "in assessing tenders with innovative content, the use of Most Economically Advantageous Tender (MEAT) criteria allows combinations of whole-life costs and quality to be assessed, increasing the chance of selecting an innovative outcome. Risk aversion is a particular problem in the public sector, especially when benefits go beyond the electoral horizon. However, risk can be effectively managed and mitigated, with partnership an important potential solution."

Norway has already garnered some experience in supporting innovative public procurement (*e.g.* the OFU programme, see section 3.5.2.4 in Chapter 3) which may provide a good basis for intensifying the use of procurement for the purpose of stimulating innovation. The Ministry of Trade and Industry recently invited industry leaders, organisations (such as NHO and LO) and researchers to a broad meeting to discuss the usefulness of including innovation in public procurement and ways to do so.

1.3.8. Concluding remarks and policy considerations

The main conclusions to be drawn from the above are that framework conditions and policies in Norway are currently at least adequate to support a high level of innovation activity, indeed a level that is higher than is actually measured by the usual indicators, especially but not only, of R&D spending and IPR activity. In particular, the Norwegian labour force has a high level of education across most ages and for both men and women (including in scientific and technical skills), and there is for the moment no fundamental shortage of persons with the appropriate research skills. It is true that there has been a tendency towards weakening educational performance, and, as in many other OECD countries, there has recently been a substantial drop in the numbers of students opting for scientific and technical disciplines. The numbers in Norway are particularly low. Educational policies have been modified to address the problem, and while there is still a long way to go, the early signs are positive.

Another positive labour-market feature is that it is sufficiently flexible for new processes and products to be introduced without too much disruption. There seems to be no ingrained resistance to change in workplace conditions, but rather an acceptance that there is no progress without change, and that innovation benefits society as a whole, not just the innovating enterprise. It is not particularly difficult or expensive to lay off workers for economic reasons, and the unemployed – including older unemployed – normally find new jobs relatively quickly, if they are motivated to do so.

Other positive features are that the economy is reasonably stable at the macroeconomic level, indeed surprisingly so, given the substantial positive (and sometimes negative) shocks from the oil sector. Competition policies have been tightened and are now on the same footing as in EU countries, so firms have incentives to innovate their way out of market pressures. The financial markets are not well developed, even taking into account Norway's comparatively small size, but they are expanding in the private equity and venture capital sectors.

Finally, fostering innovation has been a priority of successive governments. Institutions exist to encourage innovation, including via information flows and fiscal help, and there is awareness at all levels of the importance of innovation for economic performance and competitiveness. Overall framework conditions appear sufficient to sustain high levels of innovation activity in the business enterprise sector.

1.4. Assessing the efficiency of the innovation system: the methodological approach

1.4.1. The concept of an innovation system

Defining and describing innovation systems is not an academic concern. It has major implications for the balance and mix of policies needed to improve innovation system performance and for the amount of communication and co-ordination required to create holistic innovation policies. To the extent that countries operate with a narrow "innovation system map" focused on science and technology and the formal R&D system, they are likely to be guided into making policy choices that optimise the formal part of the system at the expense of the whole. During the current decade, a broader perspective on innovation systems is underpinning attempts by governments to develop holistic innovation and research policies, as has certainly been attempted in the Nordic countries (Arnold *et al.*, 2006) and strongly advocated in the UK (NESTA, 2006).

Achieving an appropriate balance among systems components and policies requires adequate resources and governance, mixing co-ordination and strategy with the flexibility and receptiveness to bottom-up initiatives necessary to adjust to changing realities. Policy makers and other stakeholders need continuous information about the innovation system; they also need to develop open systems able to reflect on performance (for example through evaluation) and to consider future opportunities. Hence, there is a need for information – strategic intelligence – about the system as a basis for making policy. Some of this comes through the normal activities of actors such as innovation agencies, which collect information and experience as they work. Other intelligence comes from dedicated studies and special exercises such as foresight and planning. Evaluation provides a significant feedback loop. These elements taken together provide the information basis for policy learning, while the ability of the system and its component actors to make use of the information and to adapt and improve policies over time depends not only on the availability of information but, crucially, on having institutions and governance arrangements in place that can effectively connect knowledge to policy practice.

The innovation systems concept is allied to a number of important ideas – specifically about how innovation functions and who is involved – which are essential to an understanding of such systems.

Box 1.6. The evolution of the national innovation systems concept

Christopher Freeman introduced the term "innovation system" into the literature in a study that aimed to explain and learn from the success of Japanese research and innovation policy (Freeman, 1987). At that point, Freeman's definition was rather narrow: he referred to the state institutions involved in defining and performing research and innovation policy. Subsequent work collected in Nelson (1988, 1993) and Lundvall (1988, 1992) widened the definition of national innovation systems considerably to include industry and more of the national context within which research and innovation took place. Lundvall's perspective (inspired by the highly networked SME structure of Danish industry) focused on the interactions between business enterprises as users and producers of innovative technologies. Business enterprises were therefore put at the centre of the innovation system, although the importance of wider cultural and macro-system environments was also highlighted.

Subsequent studies (*e.g.* Metcalfe, 1995; OECD, 1999; OECD, 2002), define a national system of innovation as a set of distinct organisations (*e.g.* firms, research institutes, universities) which jointly and individually contribute to the development and diffusion of new technologies. They do so within a wider set of institutions and social, economic and political conditions that influence the organisational actors and provide the framework within which governments form and implement policies to influence the innovation process. It is, therefore, a system of interconnected organisations or core actors and wider framework conditions within which societies create, store and transfer the knowledge, skills and artefacts which contribute to innovation. From this perspective, the innovative performance of an economy depends not only on how individual organisations perform in isolation, but also on how they interact with each other and their interplay with social institutions such as values, norms and legal frameworks (Smith, 1996). In effect, each component of the system needs to work at least at an acceptable level of quality and efficiency and the linkages between them need to function well.

Interconnection and interdependence are at the heart of the innovation system concept. The innovation systems perspective originated in deliberate opposition to simpler, more or less monocausal views of innovation and the economy. Modern models of the innovation process are complex, with many linkages among actors (Mowery and Rosenberg, 1978; Kline and Rosenberg, 1986). Innovation processes do not always start at one particular place (basic science or the market) but can be prompted by changes anywhere.

Innovative activity encompasses a wide range of phenomena. Innovation systems are not concerned solely with the types of innovation that are globally novel. A lot of the strength of the Norwegian innovation system comes from its effective use of technologies developed elsewhere. It is now recognised that important forms of innovative activity include changes that are new to particular industries or individual firms. Innovation also encompasses not only "hard" technological innovations, but also softer forms concerned with organisational arrangements and procedures. Norway, for example, is strong in some of these areas, especially those that are enabled by harmonious labour relations in an economy with high employment, skills and easy labour mobility.

Innovation activities are much more than R&D. Discussions about the core scientific and technological functions in national innovation systems often jump quickly from "science and technology" to "research and development". Consequently, maps of the R&D system easily become taken as maps of the innovation system. This tends to be reinforced by heavy reliance on data on R&D inputs and outputs as indicators of the main features of innovation systems. This seriously distorts the picture because it leaves out many other kinds of S&T activity that play important, central roles in innovation.

Design, engineering and management play key roles in innovation systems. The core activity at the heart of almost all innovation is the creation of a set of specifications (or designs) of the change that is to be brought into use. These may consist of complex computer-aided designs, or specifications for procedures and organisational arrangements. In complex, one-off projects such as designing and building equipment to exploit the more difficult North Sea oil and gas fields, these skills are indispensable, and the degree of novelty involved in individual projects can mean that they are hard to distinguish from R&D skills. Indeed, in many cases, such a distinction may be, in practical terms, meaningless.

However, R&D activities may nevertheless play an important role even in this type of innovation. On the one hand, design, engineering and management may be carried out on the basis of recently developed new knowledge, perhaps even created by R&D activity from another source. In these cases they contribute to the process of translating knowledge outputs from R&D into the concrete realities of implemented innovation. Second, in addition to the "supply side" role, design, engineering and management activities play an equally important role in the other direction – from the production of goods and services to the execution of R&D. When innovators' existing knowledge base is inadequate to meet the demand for innovation, they actively "pull" on R&D to supply new knowledge.[28] Moreover, this pull on research or technological development is not simply a vague demand for innovation in general. Instead, these activities serve to concretise generalised demand into specific technical configurations or performance requirements that help to shape the process of technological development.

Business enterprises are central actors in the system. Since the earliest contributions to ideas about the innovation system, different emphases have been placed on different system components (see Box 1.6). Numerous reports have focused on public sector organisations and policy-making structures, leaving business enterprises as minor entities on the edge of system maps. In some cases, national innovation systems have been defined almost exclusively in terms of public-sector actors, quite commonly depicted within hierarchical structures through which they influence and drive other actors, including business enterprises. Other studies put business enterprises at the centre of the innovation system, and public scientific and technological organisations are somewhat peripheral. This report is based on a combination of these two perspectives. Finding the right balance between policies addressing the business sector and the knowledge infrastructure is a key task for policy makers.

Demand, not just supply, drives innovation systems. It is now common to argue that linear models of knowledge running in one direction from R&D to commercialisation provide an inadequately simplified representation of what happens in the innovation process. This model has thus been extended to include various knowledge flows running in the opposite direction (from markets to research) as these are highlighted as drivers and shapers of the innovation process. The articulation of effective demand for innovation and for knowledge and skill inputs to innovation is centrally important in stimulating or constraining innovation and the directions it takes. Policy implications include the opportunity to use

28. Indeed, R&D is often not a source of innovation but an effect of innovation decisions (Smith and West, 2005). From this perspective R&D should be seen not only as a process of discovery but also as a problem-solving activity within existing innovation processes.

procurement and regulation as well as improved supplier-user communication and partnership as ways to encourage innovation.

Innovation functions do not map tidily to organisations. Many approaches define innovation systems primarily in terms of organisations (universities, research institutes, firms, etc.). It is important to highlight that single functions rarely map to single types of organisation. Many of the key organisations in innovation systems are multifunctional; for example, universities have extended their traditional function of basic/strategic research to technology development and even further downstream to design, engineering and entrepreneurship. The functions of universities and applied research institutes increasingly overlap at the more fundamental end of the range of such institutes' activities (Arnold *et al.*, 2007). Similar functions may be undertaken in different organisations; for example, part of the process of creating scientific and technological human capital for innovation systems is carried out by specialised education and training organisations, but a very important part is also carried out by business enterprises via large expenditures on education and training and by active management of the process of experience accumulation. Mappings between functions and institutions that work in one innovation system may not be transportable to others. For example, under other circumstances, companies might do some of the applied work done by industrial institutes in the Norwegian system.

National systems are internationally open. International components of the system are increasingly diverse. For example:

- Inward flows of technology embodied in final consumer goods and services.
- Collaboration along international value chains in creating, transferring and implementing innovation in local production for export.
- The execution of local investment projects that draw on imported engineering and project management services, licensed technology and capital goods.
- Collaboration with foreign partners in scientific research or technological development.
- Inward and outward flows of FDI by multinational enterprises.
- The emigration, return and original immigration of all sorts of qualified scientific and technological human resources.
- Inward and outward flows of students.

The quantities, qualities and directions of all these flows are highly variable, and that variability has major implications for the domestic parts of the national innovation system. In many countries the active management of these international interfaces of the innovation system is increasingly seen as a major area for policy attention.

1.4.2. The government can help improve innovation system performance

Another important aspect of the innovation systems heuristic (taken over from evolutionary economics) is the idea that firms and other actors have "bounded rationality" and this – together with the idea of interdependence – makes knowledge, learning and institutions central to overall performance. Learning means there is "path dependence": what you can do tomorrow depends upon what knowledge and resources you have today and what you can do to adapt them. Interventions to improve knowledge and capabilities can change the trajectory of the innovation system and therefore its performance. Correspondingly, public support for innovation and R&D is increasingly concerned with improving participants' capabilities and promoting learning.

However, accumulated capabilities and experience can lock in parts of the system to configurations that perform badly. Unlearning as well as learning may be needed. This may require the state to play a role as an agent of change (this is in some cases an explicit task of innovation agencies) and it may require additional resources that can be used in new ways (such as Norway's Research and Innovation Fund).

The idea that market failure leads to underinvestment in research (Arrow, 1959; Nelson, 1959) has been the principal rationale for state funding of R&D since the early 1960s. In the innovation systems perspective, the presence of bottlenecks or other failures that impede the operation of the innovation system can also constitute crucial obstacles to growth and development (Arnold, 2004):

- *Capability failures.* These amount to inadequacies in potential innovators' ability to act in their own best interests. Norway has a strong tradition of intervening to help established companies develop their capabilities.
- *Institutional failures.* Failure to (re)configure institutions so that they work effectively within the innovation system.
- *Network failures.* These relate to problems in the interactions among actors in the innovation system, such as comparatively poor university-industry links.

- *Framework failures.* Effective innovation depends partly upon framework conditions, including well-functioning markets, innovation-friendly regulations etc., as well as other factors such as the level of sophistication of consumer demand, culture and social values.

The occurrence of these failures provides a rationale for public policy intervention not only through the funding of research, but more widely in ensuring that the innovation system performs as a whole. Because systems failures and performance are highly dependent upon the interplay of characteristics in individual systems, there can be no simple rule-based policy as is possible in relation to the static idea of market failure. Rather, a key role for government policy making is bottleneck analysis, which requires continuously identifying and rectifying structural imperfections.

Achieving an appropriate balance between the innovation system and policies requires adequate resources and governance, combining co-ordination and strategy with the flexibility and receptiveness to bottom-up initiatives necessary to adjust to changing realities. Policy makers and other stakeholders need continuous information about the innovation system; they also need to develop open systems to assess performance (*e.g.* through evaluation) and consider future opportunities. Hence, there is a need for information – strategic intelligence – about the system as a basis for making policy. Some of this comes through the normal activities of actors such as innovation agencies, which collect information and experience as they work. Other intelligence comes from dedicated studies and special exercises such as foresight and planning. Evaluation provides a significant feedback loop. These elements taken together provide the information basis for policy learning, while the ability of the system and its component actors to make use of the information and to adapt and improve policies over time depends not only on the availability of information but, crucially, on having institutions and governance arrangements in place that can effectively connect knowledge to policy practice.

Chapter 2

INNOVATION ACTORS IN NORWAY

2.1. Introduction

This chapter describes the key players and processes in Norway's innovation system. It focuses on the actors performing R&D and innovation activities, mainly the business sector, the public research institutes and the universities. The interaction between these groups is briefly examined, including the supporting infrastructure. Commercialisation of public sector research is discussed, together with the financing of innovation and the human resources dimension of the national innovation system. The role of government in providing basic incentives, institutional frameworks and support measures for R&D and innovation, notably through Innovation Norway, the Norwegian Research Council and SIVA, is examined in Chapter 3.

It is important to take the country's geography, economic specialisation patterns and cultural and institutional characteristics into account when assessing the state and potential of Norway's innovation system, notably:

- *Norway's topography* is an economic asset, *e.g.* for developing tourism, shipping, aquaculture, etc., but it is challenging in several respects (*e.g.* physical infrastructure, relatively isolated communities).
- *The development of the Norwegian economy has been shaped by the exploitation of natural resources.* A long tradition in fishing has recently been complemented by a strong export-oriented aquaculture industry. The discovery and extraction of oil and gas, and the development of related industrial activities in engineering and services more broadly, have had a profound influence on the economy.

- *The important role of SMEs in R&D.* The share of SMEs in business R&D is larger than in all other OECD countries, except New Zealand (Figure 2.1). However, the discussion of the "Norwegian puzzle" (see Chapter 1) suggests that large firms, such as Statoil and Telenor, may account for a relatively higher share of (especially "unrecorded") innovation than of R&D.

Figure 2.1. Share of business R&D by size class of firms, 2003

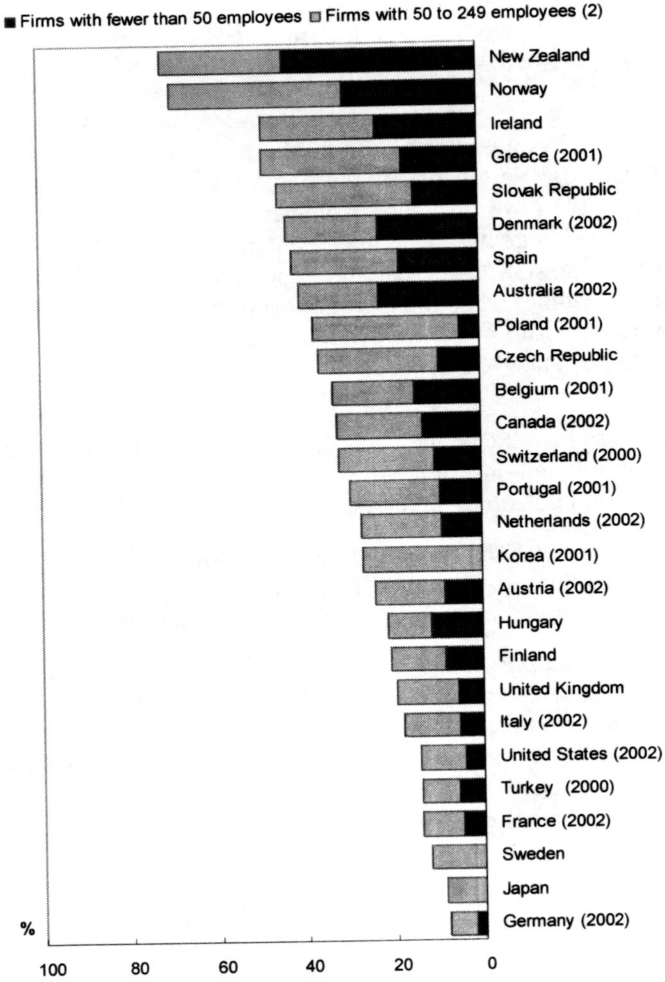

- *Cultural characteristics favour innovation processes that require a high level of social consensus on the demand side and participatory management on the supply side.* Norway shares many cultural features with the other Nordic countries, including an egalitarian society, a high degree of individualism, and relatively high tolerance for uncertainty, which, on balance, seem to be conducive to innovation on the shop floor. Management tends to be consensus-oriented and individuals are expected to take responsibility and, in turn, resist being micromanaged.
- *The labour market encourages risk taking ("flexicurity") but limits rewards (small wage disparities).* Norway also shares with other Nordic countries an organisation of the labour market that has become internationally known in its Danish version ("flexicurity"). This consists of a combination of a flexible labour market, participation of social partners in designing policy, generous arrangements for maintaining the living standards of those who are unable to work, and an active labour market policy focused on strengthening the competence of the unemployed. In addition, Nordic countries have small wage disparities.

2.2. Division of labour among main R&D performers

As Figure 2.2 indicates, real Norwegian R&D expenditure levels have been on a long upwards trend in both the public and private sectors, albeit with a modest decline in business enterprise expenditure on R&D (BERD) in recent years. This decline resulted from a substantial reduction in R&D among large companies, which was partly counterbalanced by increased R&D performance in response to the Skattefunn R&D tax incentive, which is aimed primarily at small and medium-sized enterprises (SMEs). An important feature is that the industrial and public components more or less track each other through the period, with the ratio of industrial to public expenditure rising only slowly. The most recent figures (for 2006), however, show a substantial increase of business sector R&D expenditure (6.8% in fixed prices over 2005). The strongest growth of R&D was recorded by the largest enterprises (with more than 500 employees).

Figure 2.2. Development of R&D expenditure in Norway, by source of funds

NOK millions, constant prices

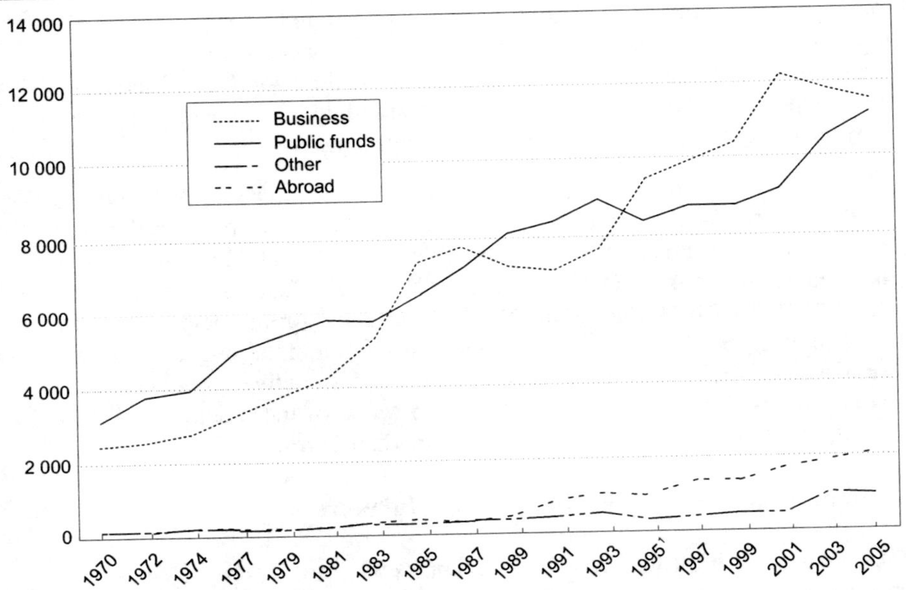

1. There is a break in the series in 1995.
Note: the scope of the survey was gradually expanded up to 1995, *inter alia* by the gradual introduction of service sector firms.

Note: The scope of the survey was gradually expanded up to 1995, by the gradual introduction of service sector firms, among others.

Source: NIFU STEP/Statistics Norway.

About 30% of all R&D in Norway takes place in the system of higher education, mainly in universities and specialised university institutions. R&D is mainly funded through the institutions' ordinary budgets, but supplementary financing is obtained for programmes and equipment, mainly from the Research Council of Norway (RCN). Some 23% of national R&D takes place in the institute sector, which obtains a large share of its income from the Research Council and foreign sources.

Figure 2.3. Development of R&D expenditure in Norway, by sector of performance

NOK millions, constant prices

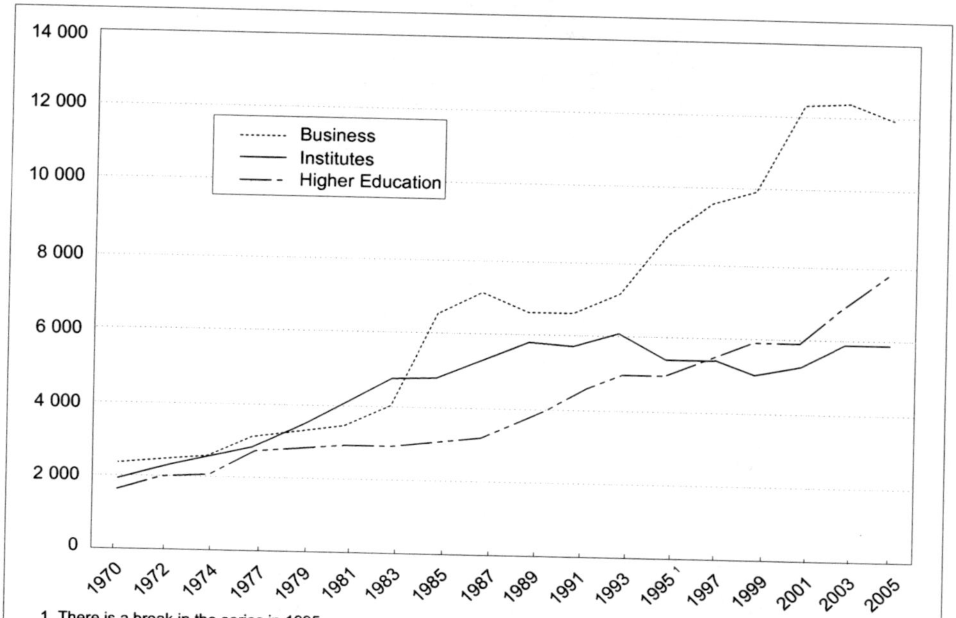

1. There is a break in the series in 1995.
Note: the scope of the survey was gradually expanded up to 1995, *inter alia* by the gradual introduction of service sector firms.

Note: The scope of the survey was gradually expanded up to 1995, by the gradual introduction of service sector firms, among others.

Source: NIFU STEP/Statistics Norway.

Figure 2.3 shows the development of R&D expenditure by sector of performance. Historically, the research institutes have been more important than higher education in Norwegian public R&D, but since 1997 expenditure in the universities has overtaken them.[29]

29. Expenditure data need to be interpreted with caution. Expenditures in the institutes are closely tied to R&D performance because they are linked to managed projects. Expenditure in higher education is largely notional: in the block grant to the universities it is assumed that academics will spend a certain share of their time on research. However, the amount of time devoted to research varies widely and no managerial process ensures that the amount of research paid for by the block grant is done.

Table 2.1. R&D in the higher education and institute sectors, 2003-05

	Expenditure NOK billions, current prices		Person years	
	2003	2005	2003	2005
Higher education sector	7.5	9.1	7 918	9 420
Institute sector	6.4	6.9	7 238	7 276

Source: NIFU STEP/Statistics Norway.

Figure 2.4. Current expenditure on R&D by field of science, 2005

NOK millions

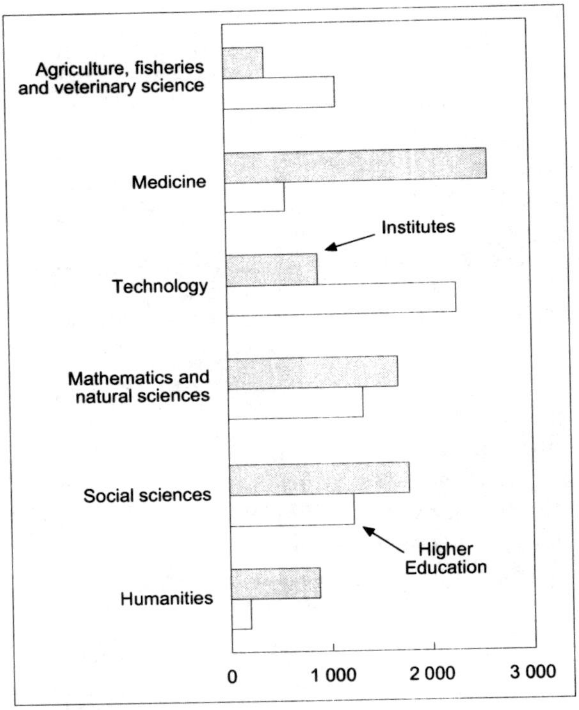

Source: NIFU STEP/Statistics Norway.

The substantial increase in the higher education sector from 2001 to 2005 is confirmed by the latest R&D statistics, which show that the R&D person-years in the higher education sector rose by 19% between 2003 and 2005 but remained stable in the institute sector (Table 2.1). Investment in buildings and equipment accounts for 10% of the 2005 figure, compared with 4% in the institutes and 7% in industry, a further indication of deliberate state investment in the higher education sector.

Unlike the countries (especially in eastern Europe) where academies of science have been the major operators of research institutes and universities have primarily been teaching institutions, Norway has a division of labour in research between the institutes and higher education. Most of the research in the humanities and medicine takes place at universities and colleges (Figure 2.4) while a large share of the social sciences research takes place in institutes. In the natural sciences, R&D expenditures are almost as large for institutes as for universities and colleges; the institutes dominate in technology, engineering and agricultural sciences.

Figure 2.5 shows that industry devotes most of its R&D effort to development. The institutes do more applied research and the universities do more basic research.

Figure 2.5. Division of labour in basic research, applied research and development, 2003

NOK millions

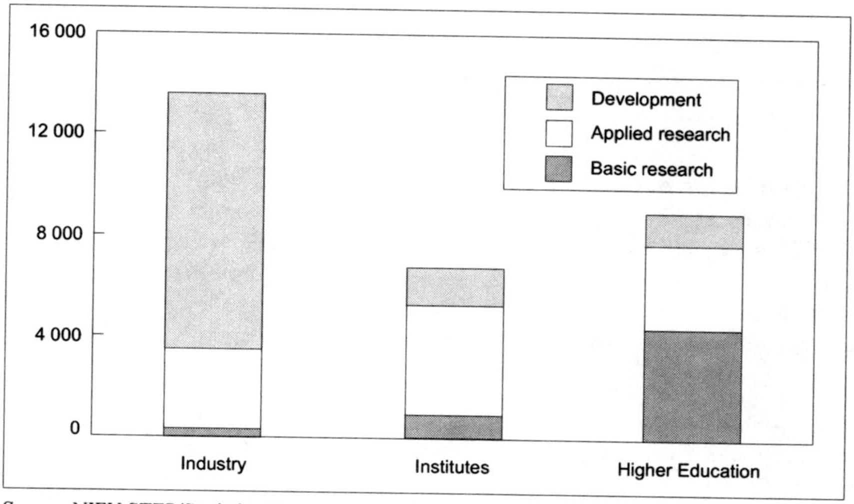

Source: NIFU STEP/Statistics Norway.

2.3. The business sector

2.3.1. Entrepreneurship[30]

2.3.1.1. Firms' size, geographical distribution and demography

Norway's business structure is characterised by a large numbers of micro firms, which represent in fact over 90% of the firm population. 28.7% (103 175) of the enterprise stock has no employees. In the private non-oil sector, there are only a few large companies. In 2005, 358 404 SMEs[31] comprised 99.6% of the stock of enterprises and employed about 63% of total employees. Their turnover represented about 55% of total enterprise turnover (Table 2.2).

Table 2.2. Enterprises, employment and turnover by firm size, 2004

Size class	Enterprises	Turnover (NOK 1 000)	Number of employees
0-9 employees	335 925 (93.3%)	1 053 424 074 (29.6%)	482 394 (30.5%)
10-49 employees	20 633 (5.7%)	639 110 466 (17.9%)	384 458 (34.4%)
50-99 employees	1 846 (0.5%)	255 390 077 (7.2%)	127 362 (8.1%)
100 and more employees	1 564 (0.4%)	1 614 746 003 (45.3%)	586 440 (37.1%)
Total	359 968	3 562 670 620	1 581 654

Source: Statistics Norway.

The distribution of the firm population by size class is quite homogenous across counties and the regional dispersion of GDP per capita is far below the OECD average (Figure 2.6). Counties situated in the southern part of Norway (Oslo, Akershus, Rogaland, Hordaland) host more enterprises than northern counties such as Finnmark, Troms and Nord-Trondelag which are more remote from the capital and more sparsely populated. The counties

30. This section draws on a report on entrepreneurship policy in Norway prepared by Fabienne Cerri and Axel Mittelstadt under the supervision of Marcos Bonturi of the Structural Policy Division (SPD) of the OECD Directorate for Science, Technology and Industry.

31. In Norway, SMEs are defined as companies with fewer than 100 employees.

situated in the southern part of Norway also host most of the large enterprises.

Figure 2.6. Regional dispersion of GDP per capita in OECD countries

Gini index, 2001

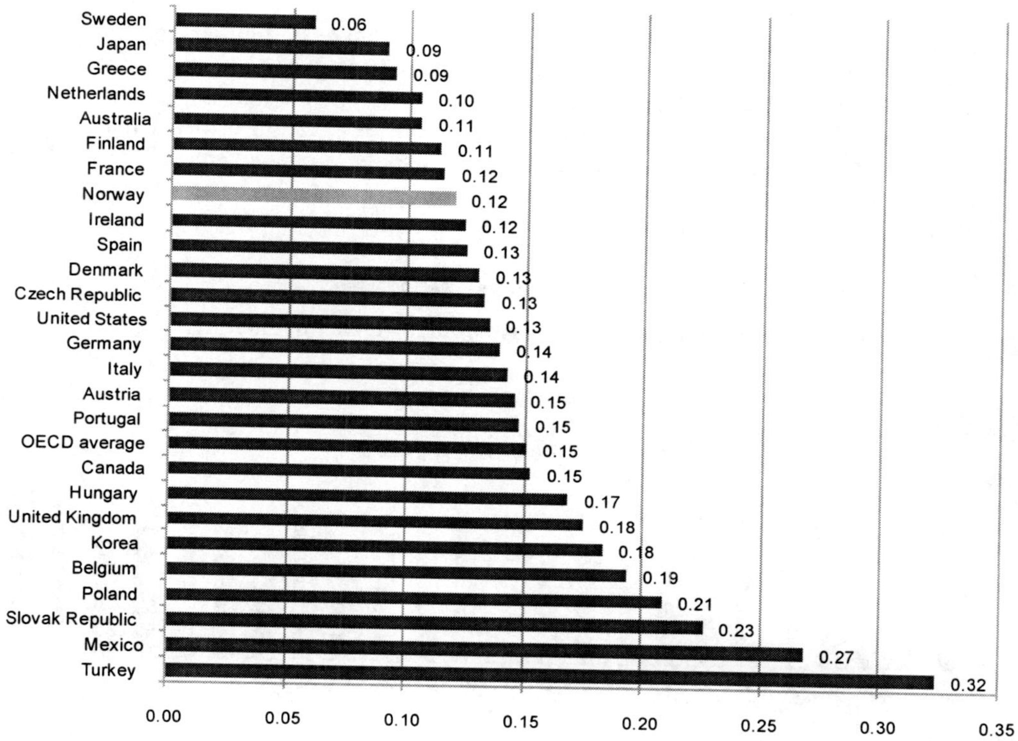

Note: 2000 data for Mexico, Norway, Poland and Turkey.
Source: OECD, *Regions at a Glance*, 2005.

The GEM international entrepreneurship survey for 2006 indicates that Norway is second only to Iceland as the most entrepreneurial country in Europe (Bosma and Harding, 2006). It indicates that 9.1% of the adult population (18-64 years) was involved in early-stage entrepreneurial activities in 2006, and 7% in 2004. Entrepreneurs in Norway are almost entirely motivated by business opportunities; few people set up companies because they do not have alternative employment. More than 60% of those involved in early-stage entrepreneurial activities wanted more independence and 35% primarily wanted more income. However, this high level of entrepreneurship is not reflected in self-employment, which accounts for only 7% of Norwegian jobs, about half the OECD average (OECD, 2007d).[32]

Rates of firm births and deaths are relatively high by international comparison (Figure 2.7). In 2006, 51 374 companies were established (17.6% of the stock), 41 121 (13% of the stock) were dismantled, and 3 032 went bankrupt (1% of the stock). The survival rate for self-run enterprises is low relative to the survival rate of other SMEs. The share of limited liability companies created in 2001 and surviving for three years was 63%, compared to self-run firms with a survival rate of only 35%. The magnitude of start-ups from 2006 partly reflects new legislation on the taxation of dividends, whereby dividends paid from one company to another are tax exempt. There are also advantages to owning property through companies. The implicit financial incentive has led to stronger firm creation.

Figure 2.7. Firm demography, 2003

Birth and death rates as a percentage of total number of enterprises

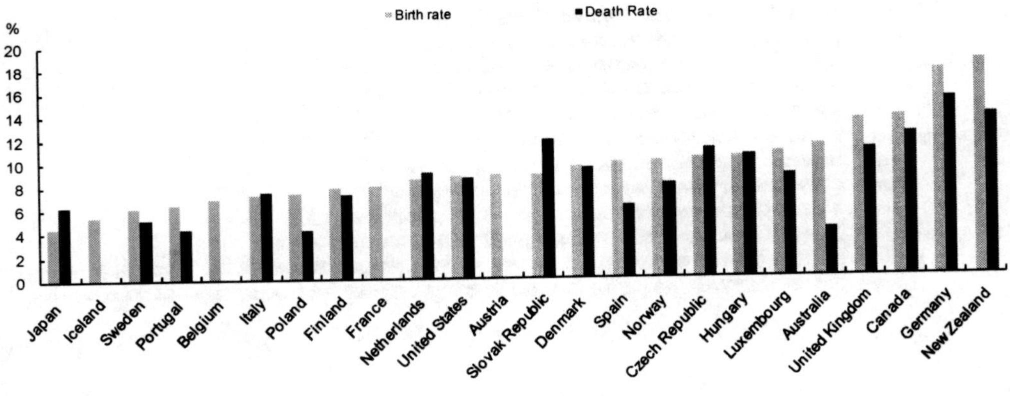

Source: OECD Science Technology and Industry Scoreboard 2007.

32. As in other countries, tax and regulatory requirements influence the choice between becoming an employee or being self-employed. Compared to employees, the self-employed have lower social security contributions but are entitled to lower social benefits. However, the self-employed may choose to establish a private limited company and employ themselves. They must then pay both the employer's national insurance contribution and the employee's social security contribution. In exchange, they receive social benefits as an employee.

According to the 2005 OECD Survey, "Policy Questionnaire on Bankruptcy",[33] possibilities for reorganising a company and chances of getting a fresh start in case of financial difficulties appear to be lower in Norway than in other OECD countries, which have early warning systems for financially distressed companies, fast-track mechanisms for re-organisation or fast-track discharge proceedings for legitimate bankruptcies. World Bank indicators from the *Doing Business* database also show that bankruptcy recovery rates are high compared to other OECD countries, indicating that the investors' potential losses are limited in case of closure.

Access to finance by Norwegian entrepreneurs seems, in international comparison, better for loans than for equity. The banking system is efficient by international standards; interest rates and risk premiums are modest, both from a historical perspective and compared to other OECD countries (Figure 2.8).[34] Overall, access to loans in the absence of collateral appears to be easier than in many other countries. This may explain the declining use of government guarantee schemes in recent years. In contrast, Norway's equity market is relatively underdeveloped (see section 2.6).

Figure 2.8. Net interest margins

Average 1996-2003

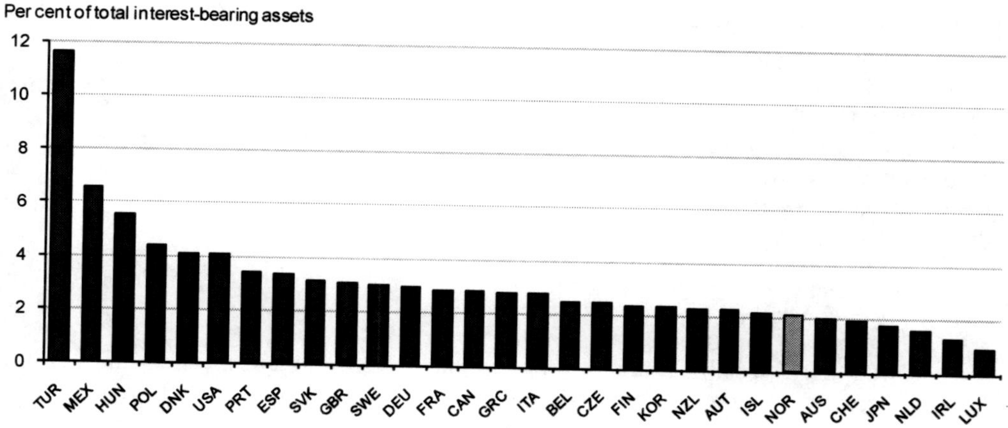

Source: OECD, *Going for Growth*, 2006.

33. The 2005 Survey on Bankruptcy analyses and compares regulation of rehabilitation, liquidation and restructuring of debt and discharge in different OECD countries.

34 It should be noted however, that direct product pricing (user payments) is rather high in Norway.

2.3.2. Business sector R&D and innovation patterns

2.3.2. R&D activities

R&D expenditure in Norwegian enterprises amounted to NOK 15.2 billion in 2006, an increase of 6.8% in fixed prices compared to 2005. The number of R&D man-years was 14 395, and 16.5% of the enterprises reported having performed R&D. Enterprises with more than 500 employees represented about two-thirds of the NOK 1.6 billion increase in business R&D from 2005 to 2006.

Manufacturing has traditionally been the main R&D performer in the business enterprises sector, spending 49.0% of gross domestic expenditure on R&D (GERD) in 2006, but service industries account for an increasing share of BERD, with 41.8% in 2006. Funding from abroad has increased in the last years and now represents around 10% of total funding.

In addition to their intramural R&D (Table 2.3), companies purchase R&D services from other enterprises, research institutes and universities. Total extramural R&D amounted to NOK 3.79 billion in 2006, a decrease of 6.5% compared with 2005. However, purchases from abroad increased by 5.8% during the same period.

When looking at aggregate national statistics, it should be kept in mind that numbers of R&D actors in the business sector tend to be rather modest in a small country like Norway. Narula (2002) points out that of the 600 or so firms reporting to be undertaking R&D, only about 50 spent over NOK 10 million or had more than ten R&D employees, and of these firms 15 were foreign-owned. He surveyed the R&D heads of 26 of the 35 Norwegian-owned firms, which accounted, collectively, for 63% of BERD. Narula divided his sample into:

- Group A: Traditional industries and formerly protected firms in traditional, raw materials-based sectors. Many of these are "national champions", in which the state has an ownership interest.
- Group B: Specialised and technology-intensive companies, which are generally more science-based, though quite few have evolved from suppliers to Group A to more diversified and comparatively technology-based activities and higher value added.

Table 2.3. Business intramural R&D, 2006

Sector	NOK millions	%
Fish farming	224.4	1.5
Oil and gas production	918.3	6.0
Total fish, oil and gas	**1142.7**	**7.5**
Mining and quarrying	36.9	0.2
Food	547.7	3.6
Textiles and equipment	59.4	0.4
Clothing	13.6	0.1
Travel goods, leather preparation	3.1	0.0
Wood and wooden products	57.5	0.4
Pulp and paper	174.5	1.1
Publishing and graphic industry	92.3	0.6
Chemicals	1263.6	8.3
Rubber and plastics	81.5	0.5
Non-metallic minerals	88.7	0.6
Metals	382.7	2.5
Metal goods	186.1	1.2
Machinery and equipment n.e.c.	1537.7	10.1
Office equipment and computers	17.6	0.1
Other electrical	314.1	2.0
Communications equipment	855.4	5.6
Medical and precision instruments	804.4	5.3
Vehicles and components	265.8	1.8
Other transport equipment	526.3	3.5
Other manufacturing (including furniture)	89.0	0.6
Recycling	7.2	0.0
Total industry and mining	**7 405.1**	**48.6**
Electricity and water supply	61.8	0.4
Construction	210.6	1.4
Total infrastructure	**272.4**	**1.8**
Wholesaling	499.7	3.3
Land and pipe transport	20.5	0.1
Sea transport	25.3	0.2
Air transport	35.7	0.2
Transport-related services	45.9	0.3
Telecommunications	718.5	4.7
Financial services	621.3	4.1
Insurance and pensions	70.4	0.5
Support services, financial services	277.2	1.8
Data processing	2 576.1	16.9
R&D	289.8	1.9
Technical testing and consultancy	1 236.7	8.1
Total services	**6 417.1**	**42.1**
Business total	**15 237.4**	**100.0**

Source: Statistics Norway.

Group A had 52 000 employees worldwide while Group B had 10 300. However, Group A performed only 65% of the R&D in the entire sample, indicating that Group B companies were much more R&D-intensive. Across the sample, only 43% of the R&D spending was undertaken in companies doing NOK 10 million or more of R&D a year, so it is clear that even among the largest R&D performers in Norway, R&D activity (at least as conventionally counted) is small and fragmented. Further, only 9% of Group A firms' R&D was done outside Norway, compared with 35% for Group B. While Narula's observations are about seven years old and the situation may have evolved, his picture of BERD by Norwegian-owned firms as fragmented and inward-looking is striking for sectors in which foreign investment does not play a significant role in connecting Norwegian research to the global innovation system.

Narula (2002) and OECD (2005a) also point to the close, almost symbiotic, relations between much of Norway's R&D-performing industry and key state institutions. Narula argues that RCN's use of programme committees which include many key industrial figures means that funding for R&D-based innovation largely goes to the big companies and tends to be locked into existing technological trajectories. He also points out the close relations between the major companies and the Norwegian University of Science and Technology (NTNU)/SINTEF dyad in Trondheim, observing that 70% of the R&D managers he interviewed (and the majority of their colleagues in R&D) were NTNU graduates. His sample of firms used 15% of their aggregate budget for external R&D at NTNU and a further 60% at SINTEF, leaving only 25% of their external expenditure for other parts of Norway and the rest of the world. Both Narula (2002) and Emblem (1995) refer to a policy of using Norwegian industrial research institutes to a certain degree as substitutes for company-internal R&D and argue that this generates scale, efficiency and the ability to share scarce national R&D resource.

This report recognises in a subsequent section that NTNU and SINTEF are in many important respects strong and valuable contributors to the Norwegian innovation system, but the lock-ins to individual institutions and technologies suggested by the foregoing and their promotion in policy probably need to be counterbalanced by stronger centrifugal tendencies to expose Norwegian industry to a wider range of technological change and stimuli, following the example of the Institute for Energy Technology (IFE) (see Box 2.4). In that sense, SINTEF's takeover of SI (*Sentralinstituttet* – its smaller equivalent organisation aligned with the University of Oslo) since Narula wrote risks increasing this problem of structural inertia.

2.3.2.2. *Innovation behaviour and performance*

The 2002-04 Innovation Survey is consistent with the R&D survey and suggests reduced activity compared with the previous comprehensive innovation survey (1999-2001). Manufacturing and service enterprises are generally stable in terms of their innovation activity. Within other industries, including extraction of oil and natural gas as well as electricity, gas and water supply and construction, there has been a decline.

Overall, slightly more than one in five enterprises has introduced product or process innovations, according to the 2006 Innovation Survey. In the business sector as a whole, 60% of enterprises with more than 500 employees introduced new or significantly improved products or processes, compared with only 20% of enterprises with 10-19 employees. The size factor is particularly evident in the mining and manufacturing industries, where 84% of the largest enterprises are innovative compared to 24% of the smallest. The difference is slightly less pronounced in the service industries.

Table 2.4. Percentage of innovating firms in the Norwegian business sector

	2001		2004		2006	
	Product and/or process innovation	Product innovation new to the market	Product or process innovation	Product innovation new to the market	Product and/or process innovation	Product innovation new to the market
All industries	29	12	26	11	21	9
10-19 employees	21	9	20	8	20	10
20-49 employees	33	13	28	12	25	10
50-99 employees	34	12	39	14	37	17
100-199 employees	47	18	44	17	40	14
200-499 employees	50	20	49	18	52	21
Over 500 employees	64	27	62	31	60	29

Source: Statistics Norway, Norwegian Innovation Surveys 2001, 2004 and 2006.

Figure 2.9. Innovative firms by degree of innovativeness, 2006

Percentage

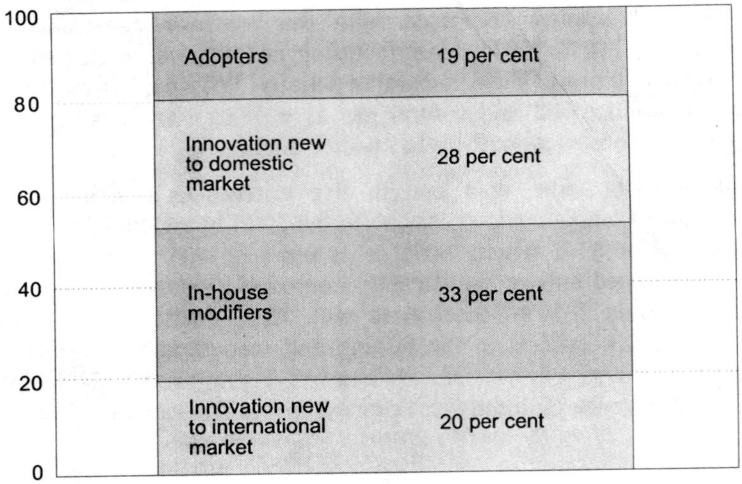

Source: SSB, Norwegian Innovation Survey 2006.

The most original innovators – those that introduced innovations that were new to an international or domestic market – make up less than half of the enterprises with innovation activities. 20% introduced innovations that were new to the international market, while 28% introduced innovations that were new to the domestic market (Figure 2.9). The least original innovators accounted for 19%.

Most innovative enterprises stated that they mainly developed their own innovations. More than two-thirds of product innovators were particularly self-reliant: While about one-quarter of product innovators developed their innovation in co-operation with others, less than 10% let other parties develop the products. Among the process innovators the differences are less pronounced, but here also self-reliant enterprises are the majority.

Manufacturing enterprises are more innovative than service enterprises. While around one-third of the manufacturing enterprises introduced new or improved products or processes, only about one-quarter of the service enterprises did so. Some industries stand out as particularly innovative, and these were more innovative than the average in past surveys as well. The most innovation-intensive industry is manufacture of chemicals and chemical products, in which 68% of enterprises are innovative. There is also a high share of innovators in manufacture of radio, television and communication equipment and apparatus. Among service enterprises, computers and related activities and telecommunications stand out, with a share of innovative enterprises of 56 and 48%, respectively. In most cases, these are industries in which R&D activities are also concentrated.

Eurostat data suggest that the proportion of innovating firms in Norway was at about the EU15 average, although it has since fallen, as indicated above. However, Norwegian firms' innovation expenditures appear to be among the lowest in Europe: 1.22% of sales, compared with 2.5% in Germany and in Finland, 2.53% in France and 1.61% in the United Kingdom (Kallerud *et al.*, 2006).

In 2006, a considerable number of enterprises reported that their innovation activities were limited or hampered. Among innovative enterprises 54% reported that the high costs of innovation hampered their innovation activities. The lack of internal funding and of qualified personnel was cited as an obstacle by 43%. Relatively fewer enterprises considered the lack of external funding and internal factors such as lack of information about technology or market as substantial obstacles to their innovation activities. Nor was there a widespread view that demand uncertainties or the dominance of other enterprises significantly hampers innovation.

Only 17% of enterprises applied for a patent during the period 2002-04. Enterprises use several other methods to protect their innovations: 22% used trademarks and 12% used copyright to protect their inventions and innovations. In terms of strategic methods, 39% made use of a lead-time advantage on their competitors, while 26% relied on secrecy. There are important differences among sectors concerning the methods of protection. Patent applications were more frequent in manufacturing (21%) than in services (14%), while trademark and copyright were used more commonly in services (28 and 16%, respectively) than in manufacturing industries (20 and 10%, respectively).

2.3.2.3. The importance of non-technological, "soft" innovation

Since 2004 Norwegian innovation surveys contain more information about organisational and marketing innovation. In 2006, 29% of all enterprises with more than 20 employees have carried out organisational changes of a kind that can be called organisational innovation. Of these, three-quarters changed the management structure or the organisational structure, while half introduced new knowledge management systems.

Of all enterprises, 21% implemented marketing innovations. Of these, three-quarters found new client bases or market segments, and close to half significantly changed the design or packaging of a good or service. The survey shows that organisational and marketing innovations are primarily introduced by enterprises engaged in product or process innovation. Moreover, large enterprises are much more often engaged in these types of innovation than smaller enterprises.

More generally, over the last 30 years, high growth in GDP and purchasing power and the development of natural resource-based innovative clusters have turned a growing share of overall demand towards services; demand for knowledge-intensive services has grown rapidly in the last decade (see below). The combination of these two structural changes – the increased share of services in final demand owing to rising national income and the increased sophistication of traditional industries – contributes to the high propensity for "soft" innovation which is at the heart of the "Norwegian paradox" discussed above. The oil and aquaculture clusters (see Box 2.1) are emblematic examples of innovation in natural resource-based industries which rely on the science base and combine R&D with engineering and "soft" types of innovation in ways that are not always easily captured by conventional indicators and statistics.

Box 2.1. Innovation in the Norwegian oil and aquaculture clusters

The development of the Ormen Lange field in the Norwegian Sea is one of the largest and most demanding industrial projects ever carried out in Norway. Hydro, a Norwegian petroleum company, is the operator. The field is situated in an area of the Norwegian Sea where climatic and oceanographic conditions make this one of the world's most challenging development projects. Norwegian research and industrial centres of expertise have been engaged to find solutions to a set of challenges that had not previously arisen for oil and gas development on the Norwegian continental shelf. Together with several partners in the Ormen Lange field, Hydro is implementing a major pilot programme to test the viability of a sub-sea compressor off the Norwegian coast. This highly innovative project would eliminate the need for a conventional platform, saving billions of NOK and halving operating costs.

The Norwegian aquaculture industry is a modern, internationally competitive industry that produces high-quality food efficiently. In terms of value, aquaculture products account for almost half of Norway's total fish exports. Salmon and trout are the main species; however, efforts are under way to farm new species, such as cod, halibut, wolffish and shellfish. Industry-related research in the fishing and aquaculture sector is conducted at a high international level. More and more knowledge and expertise are required in the marine sector to improve competitive abilities and create new employment in existing and new related industries. Many opportunities linked to the better use of by-products, biotechnology and marine resources have not yet been seized. Several companies are engaged in aquaculture across the world. For example, Marine Harvest is one of the world's leading seafood companies and produces about one-third of the world's farmed salmon and trout. It is present in 20 countries and has 9 000 employees worldwide. Other major companies are Domstein, Aker Seafood and Salmar.

2.3.3. Innovation in the services sector

Service innovation is a key driver of improved productivity and profitability in the overall economy, but it has only recently received proper attention. To better understand the policy implications of the increasing role of services in innovation, the Ministry of Trade and Industry commissioned a comprehensive study of innovation in the Norwegian services sector. The report, which proposes a fresh conceptual framework for analysing the highly heterogeneous service sector, was published in 2006 (ECON-Report No. 2006-025). Its main findings are summarised in Box 2.2.

Box 2.2. Innovation in services: a typology

The ECON/MENON report proposes an interesting typology of services to help understand why and how service firms innovate and how policy affects their innovative activity. The typology defines the following service groups:

- *Problem solvers* create value by solving specific problems for their customers. These services are not very standardised. Law firms, medical doctors, engineers, architects and researchers are typical problem solvers.
- *Producers of assisting services* generate customer value by taking over time-consuming activities that are easy to standardise for firms and households. Security services and cleaning services are typical examples.
- *Producers of digital and manual distributive services* generate value by facilitating interaction between customers, for instance by selling goods and transporting commodities, passengers and information.
- *Producers of leisure services* generate value by stimulating customers' emotions, perceptions and spiritual experience. Leisure services are very heterogeneous and include activities such as sports, arts, entertainment, restaurant services and media services.

The report finds very differentiated patterns of innovation in the four main groups:

- For *problem solvers*, innovation is often the core activity owing to a strong focus on adaptation and tailor-made solutions. Innovation surveys also indicate that a relatively large share of their innovative activity is for product innovations rather than process innovations. They focus on new solutions, new diagnostic tools, analytical concepts and differentiating brands.
- Firms that produce *assisting services* aim their innovations towards process improvements. These services have a lot in common with traditional commodity production. To a large extent, process innovation in this group is linked to improved worker efficiency through standardisation, quality control and scale effects.
- Innovation among providers of *distributive services* is a question of reducing transaction costs between customers. This can be obtained through process innovations as well as new forms of distributive services, in terms both of new ways of distributing and of what is distributed. Process innovations are often linked to digitisation and automation, and often focus on a more efficient user-producer interface. Integration of logistic systems in transport is a typical example.

…/…

Box 2.2. Innovation in services: a typology *(continued)*

- While consumers normally prefer services that are predictable and of stable quality, they often seek the opposite when they consume *leisure services*. New experiences (product innovations) are thus the most important form of innovation in this group. Leisure service providers increasingly focus on the use and development of new technology, to improve products and to reach out to a larger group of customers. New technology enables such firm to multiply their services, improve their storage capacity and simplify distribution. This is particularly relevant for providers of art, entertainment and sports services. Tourism services also tend to focus on organisational innovations that link several providers together in a network.

Share of firms with product and process innovation for five service groups and manufacturing

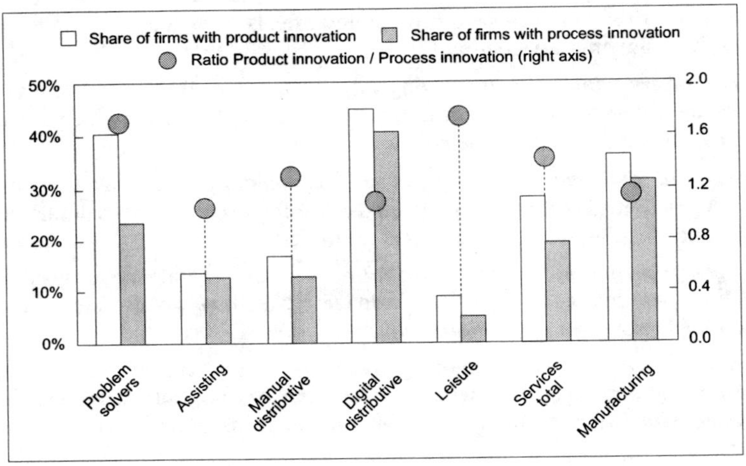

- *Source:* **Statistics Norway, CIS3 and ECON/MENON.**

Figure 2.10 shows trends in total factor productivity in selected service- and goods-producing industries.[35] Productivity growth over the last 20 years has been strongest in wholesale and retail trade, domestic transport and communication (telecommunications constitutes a large share);[36] this can be interpreted as an indirect indicator of significant innovations in trade, logistics and telecommunications, which is confirmed by anecdotal evidence (Box 2.3).

35. In 2005 ECON conducted a study of productivity growth in the banking sector which showed that since 1995 productivity growth in this sector has been higher than in other sectors of the mainland economy.

36. This contrasts with the weak growth of productivity in manufacturing, especially in the sheltered food and beverage sector and publishing industries.

Norwegian service firms are relatively R&D-intensive, compared to similar firms in other OECD countries. In terms of R&D investment as a share of value added in the services sector, only the other Nordic countries and the United States have higher R&D intensities. In 2001, service sector R&D investment represented 0.6% of GDP in Norway, while the figures for Sweden, Denmark and Finland were 1.1%, 1.1% and 0.7%, respectively. Moreover, the skill intensity of Norwegian service sector firms is high. Only Finland reports a higher share of employees with the highest level of education in this sector. The skill intensity is especially high in the financial and business services sectors.

Figure 2.10. Growth in total factor productivity in selected Norwegian industries

1983 = 100

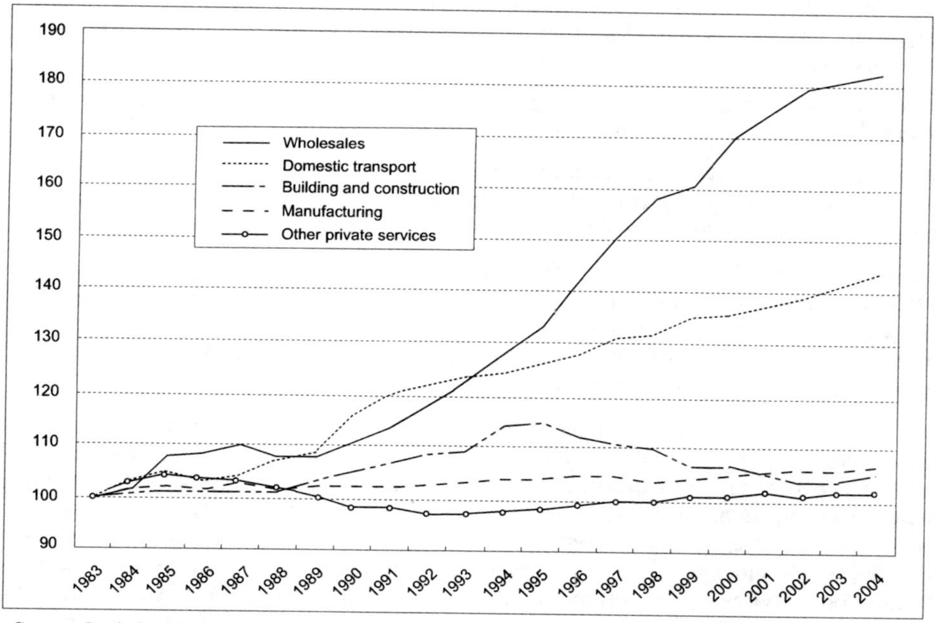

Source: Statistics Norway.

Box 2.3. Innovation in services: some Norwegian success stories

EMGS (Electromagnetic Geoservices), a spin-out from Statoil, delivers no goods, but carries out mapping services for oil and gas companies. It has developed a new method of mapping deposits of oil and gas below the ocean floor, based on known electromagnetic technology (Sea Bed Logging). EMGS's Sea Bed Logging technology significantly simplifies the search for oil.

Akvaplan-niva AS is both a consultant and a research centre for aquaculture and marine consulting. It carries out classic problem-solving services. The company is able to unite successfully "management consulting" and research. It works closely with various public institutions and is partly financed by public research projects.

Rengjøring & Vedlikehold AS is among the five largest cleaning services companies in Norway. This is a growing market with fierce competition. The firm markets itself as a quality cleaner with high focus on documented quality routines and well organised and healthy working conditions. Its innovations have mainly been process-oriented – a continuous effort to enhance the quality, train the workforce and certify their competence and processes.

CargoNet is a railroad carrier of goods between Norway's largest goods stations. Their effective delivery services have revitalised the railroad as a means of transport.

Deli de Luca is a cross between a convenience store and a café. Its key innovation is a business concept focused on high-quality convenience food combined with a refined urban atmosphere in terms of menu and location. Continuous introduction of new products is part of the concept.

SkiStar owns and operates ski facilities in Sweden and Norway. SkiStar has implemented an innovative business model. It owns ski base activities and co-ordinates complementary parts of a consistent ski experience, such as hotels, restaurants and transport, along with marketing, competence and logistics, all of which are essential for successful tourism business. SkiStar has enhanced the competitiveness of the local tourist industry.

Telenor ASA has been a leading actor in the rapid development of Norwegian telecommunication services. It was a public enterprise until 1994, when it was listed as a public corporation as part of the deregulation of the telecommunication market. Since then, Telenor has undergone a radical transformation and strong international expansion. Its recent expansion has been based on leading-edge expertise, acquired in the Norwegian and Nordic markets. Today, Telenor is among the world's largest and fastest-growing providers of mobile communications services and is listed on the Oslo Stock Exchange and NASDAQ. At the end of 2006, Telenor provided mobile communications to more than 115 million customers in 13 countries. Telenor is also the largest provider of TV services in the Nordic region and has a strong position in the growing Scandinavian market for broadband services.

Source: The ECON/MENON Report.

The picture should be nuanced if one considers growth in Norwegian service sector R&D spending over time. On the one hand, from 1995 to 2005, it grew by 146% compared to 85% for total R&D expenditure. On the other, even though overall R&D intensity is high, the share of service firms that innovate remains relatively low. According to the 2001 Community Innovation Survey (CIS3) approximately 30% of Norwegian service firms were innovative, compared to 57% in Germany and 46% in Sweden (Lucking, 2004).

Slightly under 20% of all Norwegian service sector firms covered by the CIS3 reported that they had received public R&D support, a smaller proportion than in the manufacturing sector but as much as in most other OECD countries (OECD, 2005b).

2.4. The public research sector

2.4.1. Universities

Historically, most of the Norwegian university system developed later than in other western European countries. Efforts have recently been made to upgrade regional colleges to research-performing status, helping them to act as motors of regional development. There have been significant reforms in the past few years, bringing the intellectual property regime into line with international practice, strengthening universities' commercialisation capabilities and more stringent testing of the quality of both education and research. The ratio of block grant to competitive funding is high by international standards, but this may in part be mitigated by the introduction of a research performance-driven component into the block grant.

The University of Oslo (UiO) was founded in 1811 in what was then called Christiania, after a long campaign to persuade the government in Copenhagen (Denmark) that it was not adequate for the University of Copenhagen to serve as a national university. An agricultural college was set up at Ås outside Oslo in 1859. A national polytechnic (Norges Tekniske Høyskole, NTH) was set up in Trondheim only in 1910, and a business school only in 1936 (NHH – the Norwegian School of Economics and Business Administration) in Bergen. In 1996, NTH was merged with the much newer University of Trondheim to become NTNU (Norwegian University of Science and Technology). The university in Bergen was the second to be set up in the country in 1948 and the one in Tromsø was established in 1972 with a mission to provide both higher education and research to support development in the northern part of Norway. The regional college in Stavanger achieved university status in 2005, as did the national agricultural college, now called the Norwegian University of Life Sciences (UMB). Agder University gained its new status from 1 January 2008. Accordingly, Norway currently has seven universities.

Table 2.5. Number of students by type of educational institution, 2006

	Number of students
All institutions	211 559
All universities and specialised university institutions	87 562
All university colleges	123 997
Of which state university colleges	93 169

Source: Background report.

Table 2.6. The main universities and largest colleges in Norway

Universities	Number of students
University of Oslo	29 693
Norwegian University of Science and Technology	19 873
University of Bergen	15 558
Agder University	7 429
University of Tromsø	5 876
Norwegian University of Life Sciences	2 855
Norwegian School of Economics and Business Administration	2 576
Main university colleges	
Norwegian School of Management	18 248
Oslo University College	11 516
Sør-Trøndelag University College	6 868
Bergen University College	6 201
Hedmark University College	6 001
Telemark University College	5 099
Nord-Trøndelag University College	4 894
Bodø University College	4 262
Østfold University College	4 142

Source: Background report.

Table 2.5 shows that about 41% of students are at a university; the rest are at university colleges (only the largest are listed in Table 2.6). National policy is for the university colleges to become motors of regional development and to upgrade their research capabilities. However, their funding remains primarily oriented towards teaching and the research funding programmes RCN has dedicated to their needs are very small, compared with the number of teachers who need to upgrade their skills to perform research. Experience from the KK Foundation's funding of small-scale competence centres in Swedish regional universities and colleges[37] underlines how difficult it is to make the transition from teaching to a research-based university, even when "ring-fenced" transitional funding is available.

The Bernt Commission (2001) promoted the idea that universities and colleges should more actively commercialise their inventions, and this led to a Bayh-Dole shift in the law in 2002, which abolished the so-called teachers' exception that gave university researchers rights over their inventions, assigned intellectual property rights (IPR) to the institutions, and gave the universities a more explicit commercialisation mission. As a result, the universities have set up (Birkeland Innovation at the University of Oslo) or expanded (UNIFOB at the University of Bergen) commercialisation activities, drawing on good foreign practice. At this stage, there is only limited evidence on which to base an assessment of the performance of these structures. International experience suggests that it will take many years for such structures to become viable, so it is pointless to look for short-term returns. However, a recent evaluation of the instruments and agents in place to promote commercialisation of publicly funded research is rather encouraging but notes some room for incremental improvements.[38]

Norwegian universities have been subject to reform pressures similar to those in many European countries and substantial changes have taken place in tertiary education the last ten years, mainly aimed at encouraging institutions to be more responsive to the needs of society and the economy. Following the recommendations of the Mjøs Committee, the Education Ministry launched a "quality reform" in 2003 that shortened and restructured degrees and established the NOKUT agency for quality assurance in higher education. As a result of the reform, all higher education institutions have significantly greater autonomy for managing and organising their activities. The increased institutional autonomy and accountability has led to radical

37. The KK Foundation has invested more than SEK 6 billion since its creation in 1994 in establishing research environments with distinctive profiles at Sweden's new universities and other higher education institutions, see http://www.kks.se.

38 . Einar Rasmussen, Roger Sørheim and Øystein Widding, *Gjennomgang av virkemidler for kommersialisering av forskningsresultater*, see: www.hibo.no/neted/upload/attachment/site/group1/Kommersialisering.pdf.

changes in methods and tools for higher education institutions and the authorities. There has been a transition from ministry micromanagement to management by objectives so that the Ministry of Education and Research identifies goals and monitors the institutions' goal achievement.

The university funding system has three components.

- The basic component is based on the institution's historical budget level and varies among institutions. The basic component secures stability and predictability, and represents on average about 60% of the grant to the institution. Changes in the basic component are based not on results, but on political priorities.
- The education component is an average of about 25% of the grant to institutions which is distributed on basis of the number of study points (ECTS credits) obtained by students at the institutions and the number of incoming and outgoing exchange students. This component has no upper limit and was created to give an incentive to achieve these institutions' first main goal: universities and university colleges should offer education of high international quality. Rewarding the number of study points can create an incentive to increase student throughput. Institutions can for instance increase quality through closer follow-up of students and more contact between teachers and students. The teaching component covers around 40% of the costs of a study programme, and 60% of the costs are covered through the basic component.
- Around 15% of the grant to institutions is distributed through the research component. This component was created in accordance with the institutions' second main objective: universities and university colleges should obtain results of high international quality in R&D. The research component has a strategic part and a performance-based part. The strategic part includes specific funds for PhD positions and scientific equipment, among others. The total performance-based part is a fixed amount and is redistributed among institutions each year. Four indicators determine the redistribution: number of PhD candidates, EU research grants, research grants from the Research Council of Norway and scientific publishing. The Norwegian Association of Higher Education Institutions is responsible for creating and updating a list of national and international scientific journals of high quality to be used for the last of these indicators.

A further objective focuses on the institution's ability to co-operate with external institutions, communicate research results, and meet the needs of society. The Education Ministry is currently considering an additional performance-based component in the funding system based on indicators measuring communication of R&D results and innovation. For now these indicators will not be used for the funding system, although communication, innovation and entrepreneurship play a central role in discussions between the institutions and the ministry.

2.4.2. Research institutes

Norway is a country with a comparatively large research institute sector, especially the technical research institutes on which this section focuses. These applied research institutes have long had a mission to support industrial development and continue to play an important role in supporting innovation in Norwegian industry, particularly through applied research. About a quarter of Norway's R&D activity is done in the research institute sector, about the same proportion as the university and college sector. Research grew faster in institutes than in higher education during the 1980s. However, since the end of the 1980s, the volume of institute research has stagnated and been overtaken by growth in the higher education sector. Institute staff is highly qualified, and institutes collectively employ 0.4 PhDs per full time equivalent researcher.

There are no entirely reliable internationally comparable data on the role of research institutes in the innovation system. Conventionally, their activities are classified under GOVERD[39]and the ratio between HERD[40] and GOVERD is the best internationally valid proxy for the university/institute division of labour. Different countries allocate very different proportions of state-funded research effort between the university and institute sectors. Figure 2.11 shows this ratio for a selection of countries and indicates that Norway is close to the OECD and EU15 averages. The Norwegian, Dutch and Finnish pattern is to maintain a large applied research institute system. With the exception of the Baltic countries, new member countries from eastern Europe tend to have higher proportions of GOVERD than HERD, reflecting past practice (the national academy of sciences performed much of the basic research) and the comparative isolation of the research institutes from the higher education system. France and Germany, with their large institute systems, lie near the middle. While these countries retain a large scientific research institute structure, the work of the institutes has in practice been partially integrated with that of the universities through the extensive use of joint appointments, exchange of PhD students and, in France, of "*unités mixtes*" in which researchers and academics work together.

39. Government expenditure on R&D (GOVERD) refers to R&D carried out in the government sector, not R&D funded by the government sector.

40. Higher education expenditure on research and development (HERD) refers to R&D carried out in higher education institutions.

Figure 2.11. Proportion of state expenditure on R&D allocated to HERD and GOVERD, 2005

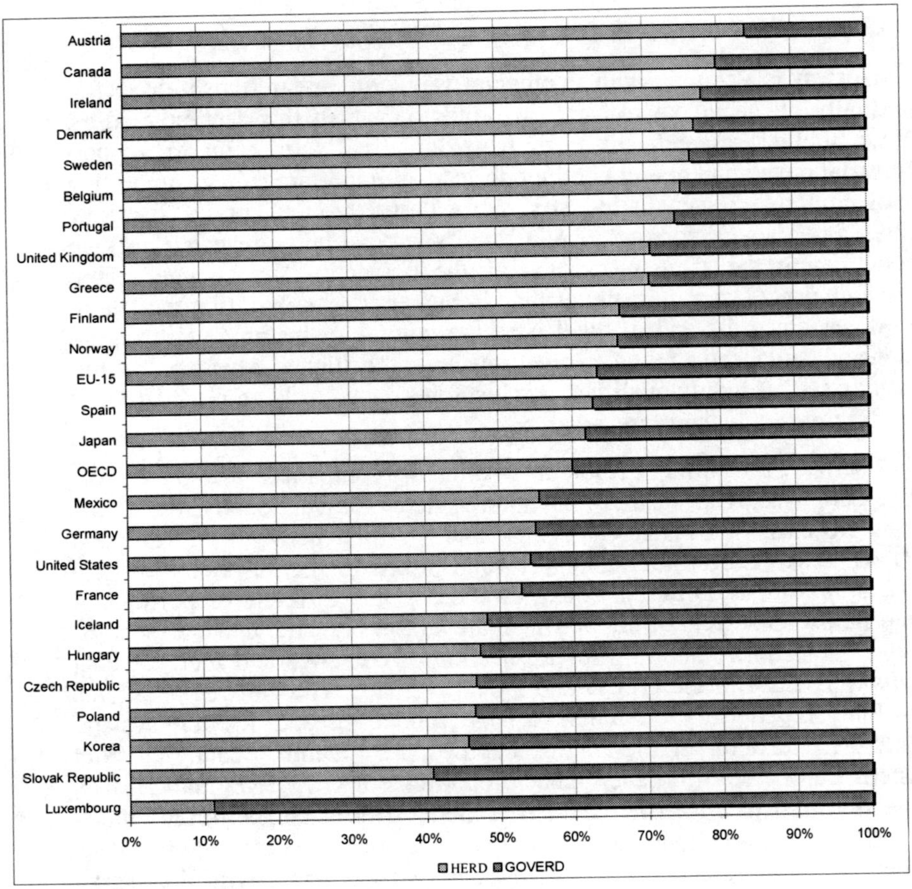

Source: OECD Main Science and Technology Indicators 2007.

Within the institute system, the National Institute of Technology (originally *Statens Teknologisk Institutt*, more recently TI) focuses on SMEs. Its 200 or so employees provide a range of product and process development services, of measurement, test and certification as well as technology-related advice and training. It was originally set up in 1916 to help small firms compete more effectively and its mission is essentially unchanged.

Key institutes were established in agriculture and fisheries in the 19th and early 20th centuries, and a technology-based environment began to arise in Trondheim around the national polytechnic (NTH) from the early part of the 20th century. However, major growth in techno-industrial institutes came after World War II, as applied research institutes were founded in Oslo at and around the Central Institute (SI), and then in Trondheim, where NTH established SINTEF in competition with Oslo-based activities. Institutes for applied social science grew rapidly in the 1960s and 1970s. Until the mid-1980s, these institutes were generally "owned" by ministries or by ministries' research councils. In the mid-1980s, however, as part of an international wave of separation of customers for research and research performers, the natural science and technology research council (NTNF) was encouraged to divest itself of its techno-industrial institutes. The techno-industrial institutes became separate foundations.

Emblem (1995) explains the importance of institutes in the Norwegian research and innovation infrastructure in terms of:

- Weak industrial R&D capability, which meant that the techno-industrial institutes could perform R&D on behalf of industry, especially because of their focus on applied research.
- The multidisciplinary capabilities of the institutes, which unlike the universities were able to tackle users' problems.
- The sector principle, which sees research as one policy instrument among others and institutes associated with ministries are used as "insiders" in policy development.

External funding of R&D in the institute sector by source of funds is shown in Table 2.7. The public financing of the institute sector is being revised. A new, indicator-based system for allocating core funding has been proposed and may be applied from 2009. This essentially rewards institutes for the quality of their work and the extent to which they tackle social needs. A key principle is that the greater the share of income that an institute receives from industry, the more it is exposed to the market failure that inhibits research performance and the more it should therefore be subsidised to enable it to develop capabilities that go beyond what market actors can create.

Table 2.7. Externally funded R&D expenditure in the institute sector by financial source and by field of science, 2003

NOK millions

Field of science	Industry	RCN	Other government	Other national	Abroad	Total
Humanities	26.9	19.6	112.5	0.9	4.5	164.3
Social sciences	106.1	425.4	468.0	30.1	89.6	1 119.2
Natural sciences	186.7	366.2	568.4	1.9	144.5	1 267.8
Engineering and technology	882.3	431.7	397.8	11.7	423.2	2 146.7
Medical sciences	13.4	40.3	303.4	16.1	27.2	400.3
Agricultural sciences	159.4	268.2	497.4	14.7	37.4	977.0
Total	1 374.8	1 551.4	2 347.4	75.3	726.5	6 075.3

Source: Kaloudis and Rørstad (2006).

Box 2.4. The Institute for Energy Technology

The Institute for Energy Technology (IFE) is a noteworthy success. Founded in 1948, IFE is an international research institute for energy and nuclear technology. Its mandate is to undertake research and development for the benefit of society in the energy and petroleum sector and to carry out assignments in the field of nuclear technology for the nation. IFE´s nuclear technology activities comprise all those that are directly or indirectly related to the institute´s two research reactors, in Halden and at Kjeller. IFE is now an independent foundation. With operating income of more than NOK 500 million and staff of over 500, it is larger than most other Norwegian research institutes.

The Halden reactor project (HPR) is one of the world's leading research facilities for the study of reactor fuels and materials behaviour, as well as of man-machine interaction in control rooms, areas that are critically important to the maintenance and further improvement of nuclear power plant safety.

IFE´s Halden reactor project has tripled its volume of international bilateral contracts during the past ten-year period, during which funding for nuclear research has generally declined. This confirms the high quality and relevance of the work performed at IFE/Halden. Bilateral collaboration between Russia and Norway makes significant and cost-effective contributions to improving the safety of the Leningrad Nuclear Power Plant and the Kola Nuclear Power Plant and radiation safety at other installations in the Kola Peninsula.

…/…

Box 2.4. The Institute for Energy Technology *(continued)*

IFE has given birth to a constant flow of spin-off companies. Sorted by year of establishment, these are:

- 1971 Scandpower AS
- 1991 Well Flow Dynamics AS
- 1994 Rembra AS
- 1995 Lokaldata Instruments AS
- 1996 Nordisk Energikontroll AS
- 1996 OMX Technology
- 1996 GE Healthcare AS
- 1998 Kjeller Vindteknikk AS
- 2000 Applied Petroleum Technology AS
- 2000 Querqus AS
- 2001 Norse Decom AS
- 2001 Scandpower Petroleum Technology AS
- 2003 Navita Systems AS
- 2004 Scandpower Risk Management
- 2004 Hybrid Energy AS
- 2004 New Energy Performance AS (NEPAS)
- 2005 Resman
- 2005 Wirescan AS
- 2005 Hydrogen Storage & Systems AS

Research institutes are building closer international links with universities through joint appointments and joint PhD education. As noted above, the French CNRS (National Centre for Scientific Research) directs its activities towards "unités mixtes" on university campuses, and about 80% of its researchers now work in such mixed research units. KTH in Stockholm has developed a strategy for closer integration with the research institutes located on campus. For its part, Denmark has merged much of its institute system into the universities.

To some degree, Norway has followed a similar trend. The SINTEF and NTNU boards established a common strategy in 2005, while the new university in Stavanger has become the owner of IRIS (formerly Rogaland Research). However, RCN's indicator report for 2005 also shows that, owing to the recent expansion of the higher education system, the amount of PhD supervision by institute staff has declined. In 1999, 750 institute researchers acted as PhD supervisors; in 2004 the number had fallen to 485. In other countries joint appointments and exchange of work between universities and institutes have been increasing, but have been stable in Norway for much of the last decade. In an international perspective, this is disturbing and warrants closer examination. The future success of the institutes (especially the vital applied institute sector) depends upon their ability to cope with the increasing role of science in R&D while the universities need the contact with industry and "real" problems that the institutes can provide.

Figure 2.12. Institutes' income by type of institute and income source

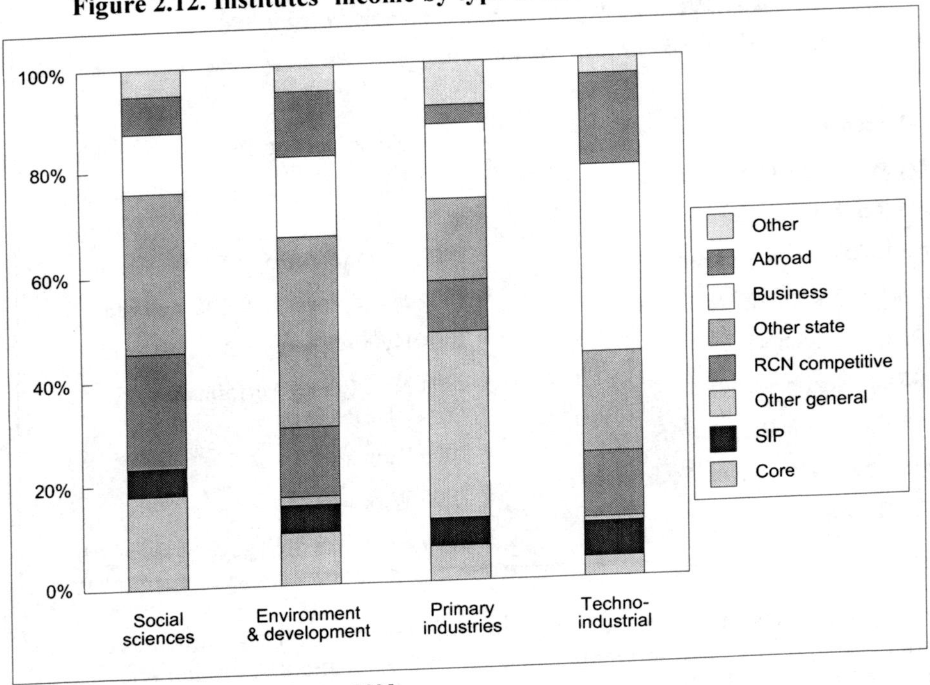

Source: RCN Annual Institutes Report, 2005.

It is important to understand how applied the Norwegian industrial institutes are. The most obvious indicator is the proportion of core funding, shown in Figure 2.12. "Core" indicates unconditional core funding. "SIP" relates to strategic institute projects, competitively awarded projects funded by RCN to build capabilities. "Other general" relates to general funding for regulatory, measurement and other tasks carried out for the state; this mainly concerns the environmental and primary (agriculture, fishing) institutes.

The international competitiveness of some of the institutes is borne out by the fact that about NOK 362 million (17%) of their industrial income in 2005 came from abroad. Overall foreign income was NOK 641 million in 2005 (the remaining NOK 279 million came from international funders, primarily the EC).

2.4.2.1. SINTEF

SINTEF is an industrial research institute and the largest of the Norwegian research institutes. Its headquarters are located on the campus of the Norwegian University of Technology (NTNU) in Trondheim. It has two main missions:

- To promote technological and other industrially oriented research at NTNU and develop co-operation between NTNU and the nation's industry and commerce, other research organisations and academic institutions.
- To meet research and development needs in the private and public sectors.

Depending how one counts, SINTEF receives core funding of 4% (unconditional funding), 7% (if RCN-funded strategic institute projects are included) or 10% (treating RCN's automatic co-funding of EU projects as core funding). A recent study of the institute sector (Arnold *et al.*, 2007) which tried to take account of synergies with NTNU,[41] suggested that the total equivalent of core funding was as high as about 20%. However core funding is counted, it is towards the low end of the international range, and this forces SINTEF (like other Norwegian industrial institutes) to work very closely with its industrial customers on projects that may involve research but may also require development and other activities on the border of R&D as it is statistically defined (Figure 2.13). For example, SINTEF plays a significant role in finding one-off engineering solutions for exploitation of the North Sea oil and gas resources. In many cases its role appears to be an extension of companies' R&D and engineering capabilities. This may be one factor in the "Norwegian puzzle". In other countries one might expect to see at least some of this work performed internally by industry and hence appearing in the statistics as BERD rather than GOVERD.

SINTEF staff is very highly qualified, with an unusually high level of PhDs (29%). Unlike some other industrial institutes abroad, SINTEF has not historically built on the measurement and testing role. Its purpose from the start has been industrial development through applied research and development, partly based on a vision that research institutes can deliver usable products and processes to industry. It describes the bulk of its work as multi-disciplinary contract research with "problems set and solved in the context of application".

41. For example, some employees work sometimes on applied research at Sintef and sometimes on basic research in NTNU.

Figure 2.13. Institutes' core funding: an international comparison, 2005

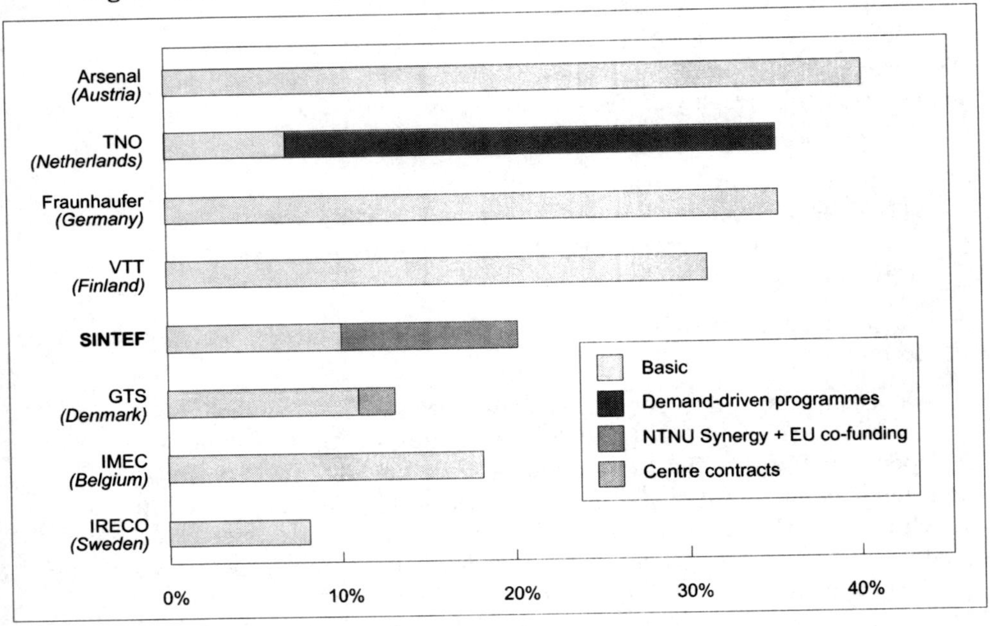

Source: Arnold *et al*, 2007.

SINTEF's key role in Norwegian industrial development means that it has longstanding relations with many major Norwegian companies, such as the Hydro group. Another large block of industrial demand comes from medium-sized firms, owing to the Research Council's user-directed R&D programmes.

A fairly recent development has been to build longer-term strategic alliances with key customers on a formal basis. For example, SINTEF signed a NOK 50 million agreement with Sydkraft, establishing a working relationship for the period 2002-06. There are nine-year agreements with two oil companies about flow assurance in oil pipelines.

SINTEF's origins and history mean that its relationship with NTH/NTNU has been close since it was founded. In 2003, 537 of SINTEF's employees held some sort of post at NTNU, in addition to their SINTEF duties. Especially in Trondheim, many facilities are shared between the university and SINTEF. In 2003, SINTEF estimated that it and NTNU both invested some NOK 30 million in shared equipment and running costs for the year. In 2004, 89 PhD students with external grants were placed at SINTEF for their studies. In addition, 32 members of SINTEF's staff were working for their doctorates, for a total of 121 PhDs in progress. In all, 109 members of

SINTEF staff were acting as PhD supervisors. There is no routine arm's length accounting of this or of the work done by one organisation that benefits the other, such as the alignment of university research and the development of technology platforms by SINTEF. Clearly, however, the relationship provides very large synergies and is highly valued by both sides. A reason why the relationship has been difficult to account for in the past has been its bottom-up character. In 2005, the Boards of NTNU and SINTEF for the first time decided to establish a common strategy.

2.5. Interaction among actors and supporting infrastructures

2.5.1. University-industry links

Establishing and strengthening links between the university system and other parts of society, especially the economy, is a policy goal in almost all countries to enable industry to make better use of existing and new knowledge, build technological capacity, focus research efforts on problems of economic and social relevance and ease the mobility of trained and educated people.

Like all innovation surveys, the Norwegian survey tends to show that universities and institutes are not the main source of the information firms use in innovation, and this is also largely true for the information embedded in patents (Table 2.8). However, the proportion of firms co-operating with the public research infrastructure, especially the research institutes, is much higher than the OECD average.

Table 2.8. Number of patents from the public R&D base

Patent applications to the Norwegian Patent Office, 1998–2003

Year	Colleges	Research institutes	Universities	Public R&D base (A)	Total number of patents (B)	Percent A/B
1998	14	38	58	110	1 291	9%
1999	9	34	82	125	1 338	9%
2000	10	54	69	133	1 406	9%
2001	13	59	80	152	1 275	12%
2002	10	71	86	167	1 267	13%
2003	9	54	48	111	1 161	10%

Source: NIFU STEP and Norwegian Patent Office.

Figure 2.14. Proportion of HERD funded by industry, 1981 and 2005

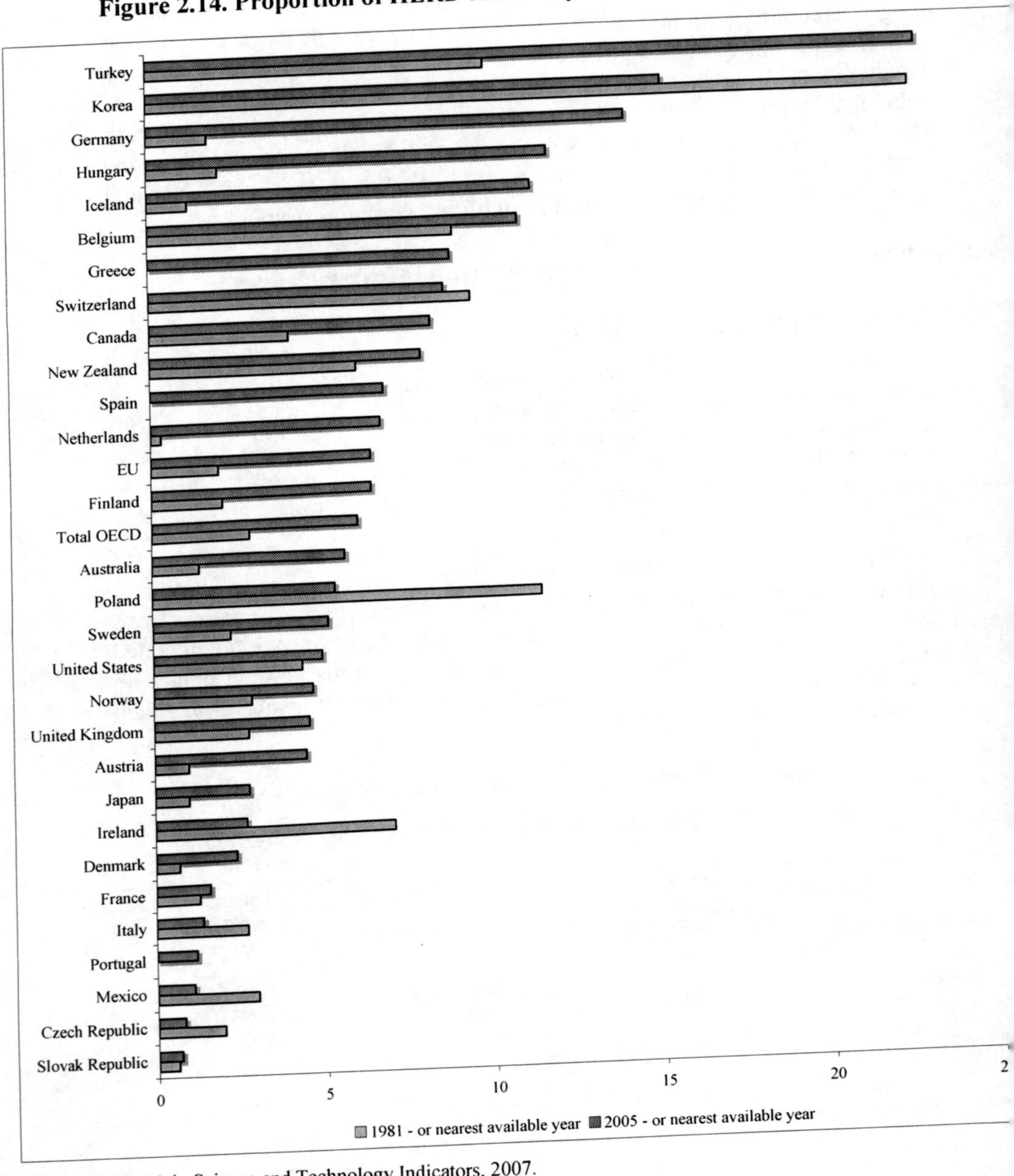

Source: OECD Main Science and Technology Indicators, 2007.

Figure 2.14 shows the proportion of HERD funded by industry in a range of countries. The proportion of HERD funded by industry in Norway is similar to that in the United States and marginally below the OECD average. These statistics are, of course, affected by an inadequate definition of HERD, (it is often a somewhat notional amount ascribed to a proportion of the block grant).

According to the 2005 RCN indicators report, industry funded roughly NOK 350 million of the research in Norwegian universities but also provided NOK 1.4 billion (22%) of the institute sector's income. Norwegian industry spent in excess of NOK 700 million on research at universities and institutes in 2004. On any measure, therefore, Norwegian data on industrial funding of HERD significantly understate the interaction between industry and the knowledge infrastructure as a whole. In Norway, as in other countries, companies work with universities and institutes for different reasons. The knowledge involved in institute interactions is normally much closer to market and may relate to a critical path in R&D, while university interactions tend to involve longer-term, more speculative knowledge, development of human capital and recruitment of skilled labour.

RCN's user-directed R&D activities are the major promoters of industrial links with the knowledge infrastructure, especially with the institutes. These have recently been complemented by the creation of a regional innovation centre scheme – the Norwegian Centres of Expertise (NCE) programme owned by Innovation Norway – and a competence centres scheme – the Centres for Research-based Innovation (CRI) (Box 2.5).

As indicated above, the universities and university colleges have been linking up with industry via technology transfer offices (TTOs) and using existing infrastructure such as incubators and science parks. These include:

- Tromsø Science Park Ltd., Tromsø.
- Trondheim Innovation Centre Ltd., Trondheim.
- Leiv Eriksson Innovation Ltd., Trondheim.
- Bergen High-Technology Center Ltd., Bergen.
- Rogaland Science Park Ltd., Stavanger.
- Campus Kjeller Ltd., near Oslo.
- Oslo Research Park Ltd., Oslo.
- Ås Science Park Ltd, Ås.

Box 2.5. The Norwegian Centres of Expertise and the Centres for Research-based Innovation

Norwegian Centres of Expertise (NCE)

This programme is targeted at specialised business clusters with profitability potential to promote the development of internationally competitive businesses and industrial centres. It was launched in 2006.

The specific content of the NCE programme was developed by the three business support agencies: Innovation Norway, SIVA and the Research Council of Norway. Co-operation and co-financing between the business support agencies, the regional partnerships and the industrial centres ensures that this programme is to be a long-term effort, strengthening both the regional and national innovation system.

The Ministry of Trade and Industry and the Ministry of Local Government and Regional Development finance the programme. In 2006, the ministries granted NOK 35 million to the six NCEs that had been appointed. Today there are nine NCEs and the 2007 budget is NOK 50 million.

Centres for Research-based Innovation (CRI)

The Centres for Research-based Innovation (CRI) scheme targets the research-intensive part of the Norwegian business sector. The main objective is to enhance the innovative capability of the business sector by forging close alliances between research-intensive enterprises and prominent research groups for long-term research projects. Participation by foreign companies or research institutions is welcome.

A centre's host institution can be a university, a university college or a research institute, or an enterprise with strong research activity. In addition to generating relevant research, increasing co-operation among various research centres and disseminating technology, the scheme emphasises the education and further qualification of researchers.

The Executive Board of the Research Council decided on 15 June 2006 that 14 of the applicants would be invited to establish centres and these are now starting up.

A centre is financed for five years, with the possibility of a three-year extension. Co-financing by government authorities and private businesses is a prerequisite. Several businesses and research institutions can be accommodated in a centre.

The total budget allocation from the Research Council for the 14 existing centres under the CRI scheme will amount to NOK 1.12 billion over an eight-year period. Each centre will receive an allocation from the Research Council of roughly NOK 10 million a year and the host institution and partners must contribute at least as much as the Research Council. Formally, the Research Council's share is financed by yields on the Fund for Research and Innovation.

The science parks have traditionally been service organisations and real estate managers. Now, however, their role as incubators and assistants for innovation is increasingly important. Many parks have their own commercialisation units or companies, and they are often local representatives for the FORNY programme.

A bridge needs a foundation at both ends, so building linkages requires technological capacity in industry as well as in the knowledge infrastructure. While there is a strong set of linkage instruments in place, there appears to be scope to strengthen the PhD cadre in industry. A business PhD scheme similar to Swedish and Danish programmes has recently been established.

2.5.2. ICT infrastructure

Norway has a good track record for investing in ICT and widely adopting and using ICT. It has developed niche expertise in firms often directly or indirectly linked with the dominant offshore, maritime, construction and infrastructure sectors (OECD, 2004). Penetration of information and communication technology (ICTs) is high and basic telecommunication connections are available at a reasonable price. As in other Nordic countries, mobile telephone charges are low (Figure 2.15). The number of broadband subscribers is among the highest in OECD countries although prices are higher than in some neighbouring countries (Figure 2.16). Use of wideband/ broadband by industry increased by 24% between 2006 and 2007. However, full use of ICT by business lags, particularly in broad sections of manufacturing. Perceived barriers include software shortcomings, the expenses linked to IT as well as the lack of flexibility among IT suppliers.[42]

42. Information from Statistics Norway.

Figure 2.15. OECD basket of low user mobile telephone charges, May 2007

Annual charge, USD PPP, including tax

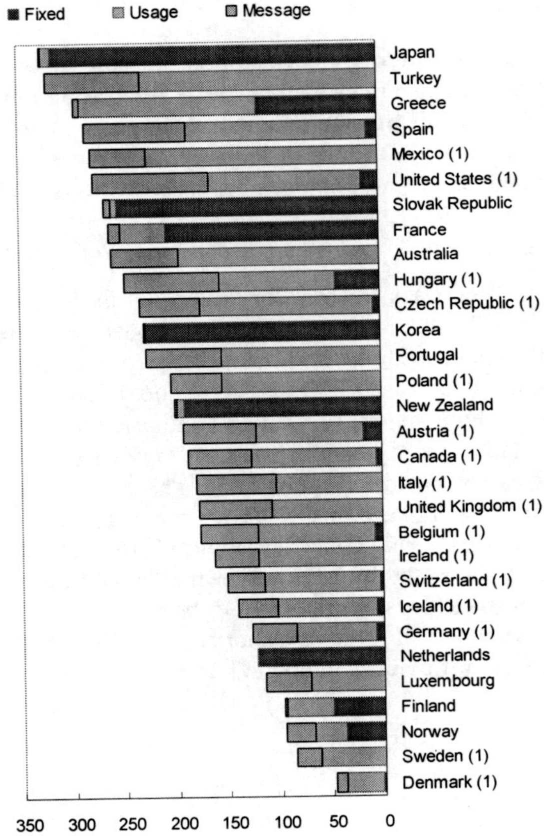

1. Package using prepaid card.

Source: OECD Science, Technology and Industry Scoreboard 2007.

Figure 2.16. Broadband prices per Mbit/s and prices for monthly subscription,[1] October 2006

USD PPP, including tax

1. Prices selected are the lowest observed price per Mbit/s and the lowest observed price for a monthly subscription which has an entry-level 256 kbit/s connection.

Source: OECD Science, Technology and Industry Scoreboard 2007.

Figure 2.17 Investment in ICT,[1] 1985-2003[2]

Percentage of non-residential gross fixed capital formation

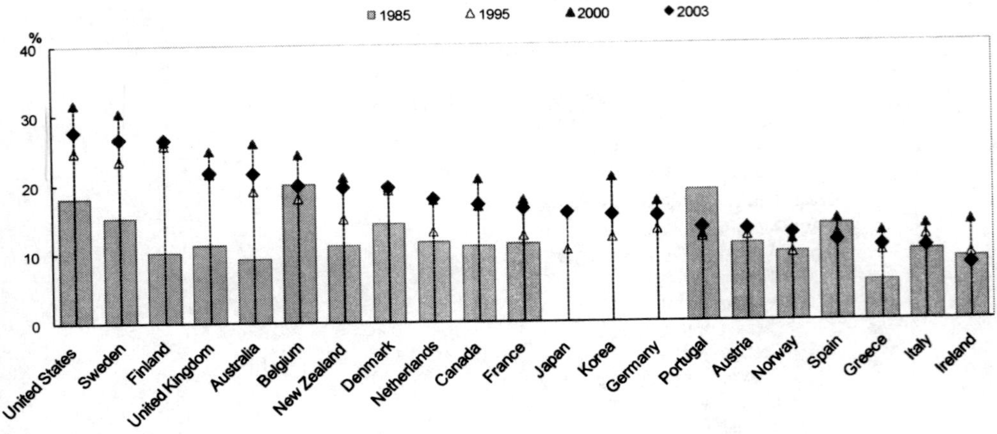

1. ICT equipment is defined here as computer and office equipment and communication equipment; software includes both purchased and own account software. Software investment in Japan is likely to be underestimated, owing to methodological differences.

2. Data for 2003 or latest year available.

Source: OECD Science, Technology and Industry Scoreboard 2007.

This may also be the result of underinvestment in ICT equipment and ICT-related research. Investment in ICT equipment and software was lower in Norway than in other OECD countries between 2000 and 2003 (Figure 2.17). Government appropriations for ICT research in the Research Council of Norway's budget have stagnated in recent years even though ICT was identified as one of four priority areas as early as 1999 and despite the conclusions of an international evaluation of Norwegian government ICT research carried out in 2002 by the RCN. This evaluation pointed in particular to the fact that the number of ICT doctorates was insufficient to satisfy academic and industrial requirements and that Norwegian ICT research was too theoretical or calculation-oriented to the detriment of technological development.

Norway ranked 12th on the Economist Intelligence Unit's 2007 eReadiness ranking. This is a drop of one place from 2006 and among the Nordic countries, only Iceland is behind Norway. Compared to 2003 when Norway ranked 4th, the drop is significant, but competition is stronger as more countries massively adopt ICT.

2.5.3. International linkages

As a small open economy, Norway's trade pattern and society are in important respects already strongly internationalised. The internationalisation of research and innovation is important both to learn from foreign research and to build research and market alliances.

Norway has a number of large companies with successful international operations but Norwegian industry's R&D appears to be insufficiently internationalised, despite commendable government efforts to assist this process, for example through the network of foreign offices administered by Innovation Norway. Multiple R&D locations increase absorptive capacity (the learning face of R&D) and provide the means to develop and offer locally adapted solutions to customers. Narula (2002) reported that not so long ago the largest Norwegian-owned R&D group outside Norway comprised only 65 people. The significance of this observation is underlined by another finding from Narula's survey of the 35 largest Norwegian R&D performers that "Norwegian firms do not, in general, consider Norway to be a useful place from which to monitor the technological activities of competitors…. Indeed, when asked to list their major competitors, none of the firms in the survey included any firms in Norway." Smaller firms are more likely to have international R&D operations.

Historically, the small scale and limited subject range of Norwegian universities meant that many people had to go abroad to study, especially at the postgraduate level. The expansion of the university and college systems in recent decades has significantly reduced the need for this, although the state's student loans and grant system continues to support study and research training abroad. Many argue that this has reduced Norwegian scientists' international networks and RCN therefore encourages PhD students to spend at least some of their time abroad in order to combat this loss (Table 2.9).

Table 2.9. Norwegian students abroad in 2005/06

	Full programme students		Visiting students		Total	
	Number	%	Number	%	Number	%
Undergraduate	7 347	57	5 149	73	12 496	62
Graduate	5 364	41	1 863	27	7 227	36
PhD	280	2	5	0	285	1
Total	12 991		7 017		20 008	

Note: "Graduate" includes students in medical school and similar integrated programmes.

Source: Background report and NIFU STEP.

There was about 9 000 foreign students in Norway in 2005 (4.3% of the total student body), a substantial increase over the level some 15 years earlier. Some 15% of foreign students stay in Norway after graduating.

In practice, Norwegian science is following the worldwide trend to internationalise, as illustrated by the dramatic growth in the proportion of Norwegian research publications now written with foreign co-authors (Figure 2.18).

Figure 2.18. Norwegian publications with and without international co-authorship, 1991-2006

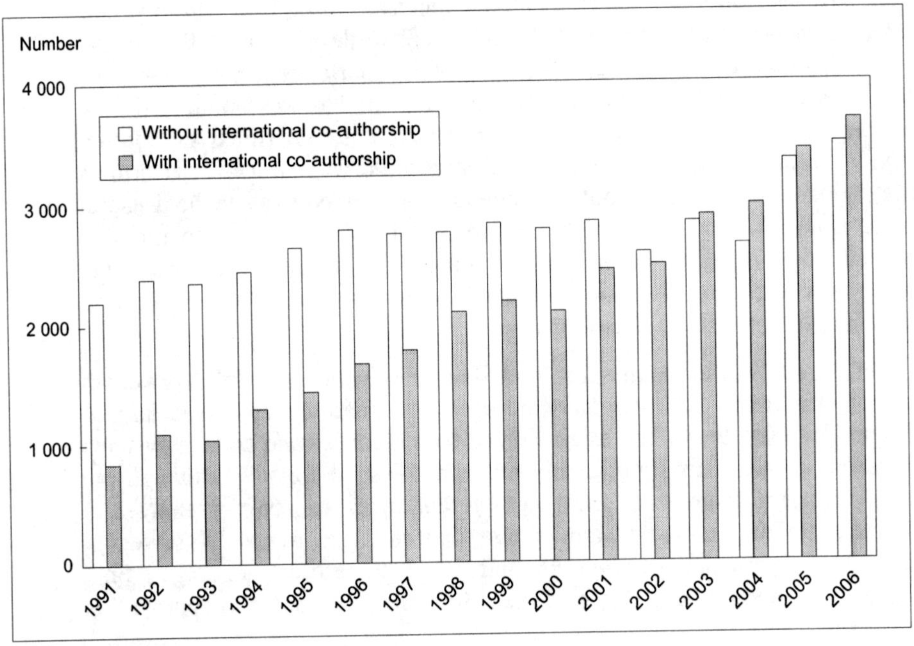

Source: Background report and NIFU STEP.

Norway is already a full participant in the EC Framework and CIS programmes, as well as engaged in most major European multilateral scientific collaborations. The evaluation of Norwegian participation in the Fifth Framework Programme (NIFU, STEP and Technopolis, 2004) found that this had useful networking effects[43] but pointed out the lack of co-ordination between the national and EU levels. RCN has been charged with providing this co-ordination. More generally, international co-operation is included as a criterion in RCN project evaluations.

43. Like most evaluations of national participation; see Arnold *et al.*, 2005.

There are extensive bilateral research agreements at the university, RCN and government levels, S&T co-operation strategy for North America was launched in 2004 and there have been recent agreements with Japan, the United States and India. The Norwegian government is currently working on establishing an agreement with China.

Nordic co-operation has for a long time been small-scale but relatively effective. Nordforsk, the joint organ of the Nordic research councils, has launched a number of small Nordic centres of excellence – virtual research centres across universities – that appear to strengthen Nordic research and support existing platforms for wider co-operation. In contrast, the Nordic innovation agency – Nordic Innovation Centre (NICe) – stands apart from national organisations and pursues its own small-scale Nordic-level agenda. Nordic energy co-operation provides a further channel for research co-operation. Recent experience with the European Commission's ERA-NETS – typically aimed at bringing together from three to six national agencies to define opportunities and implement joint calls for research proposals in areas of importance to them – has been that these involve a large effort that normally only results in a small core of co-operative funding. RCN was involved in 34 such networks, at considerable opportunity cost internally and at the cost of further fragmenting and complicating the "call for proposals landscape" to which researchers react. More focused, bottom-up efforts based on existing Nordic ties or other established relationships (for example, within the TAFTIE network of European innovation agencies) will probably be a more productive and less costly approach to joint work across borders.

While there is keen interest in internationalisation policy, the practice remains somewhat fragmented. In response to a document from the National Science and Technology Policy Council of Finland (2003), Tekes has made internationalisation part of its normal project assessment criteria and the Academy of Finland has increased its work on international collaboration in order to improve the integration of Finnish R&D communities into global networks and benefit from the fact that large Finnish firms increasingly need to do some of their R&D abroad. This suggests similar opportunities for Norway to include internationalisation in existing policies. This should start by a systematic and comprehensive evaluation of all relevant government programmes and initiatives.

2.6. Financing innovation: venture capital

The availability or lack of venture capital is a recurring theme in discussions of innovation in Norway. As in many other OECD countries, the topic tends to trigger a debate between entrepreneurs who cannot find money to support their ideas and venture capitalists who cannot find enough attractive ideas in which to invest. The strong social legitimacy of researchers means their voices are especially likely to be heard in these matters and there is therefore a widespread perception of capital shortage.

The *OECD Economic Survey: Norway* (OECD, 2007a) argues that the Norwegian equity market is comparatively underdeveloped, owing to the large role of state ownership, but that the venture capital market is larger than in many OECD countries, though well below the leading countries (Figure 2.19). The Norwegian Venture Capital Association's 2006 survey indicated that half of the capital under management was free for investment, and that this was as true in the seed capital funds as in the start-up, expansion and buyout funds. Oil and energy and ICT, as well as life sciences, finance and aqua/agricultural, receive significant amounts of funds, but generalist sectors attract most of the investment. In 2006 there were 56 management companies and 85 funds or portfolios holding a total of NOK 45 billion in capital under management. In the aggregate, therefore, there appears to be no evidence of a shortage of venture capital.

Figure 2.19. Venture capital investment flows, 2000-2003

As a percentage of GDP

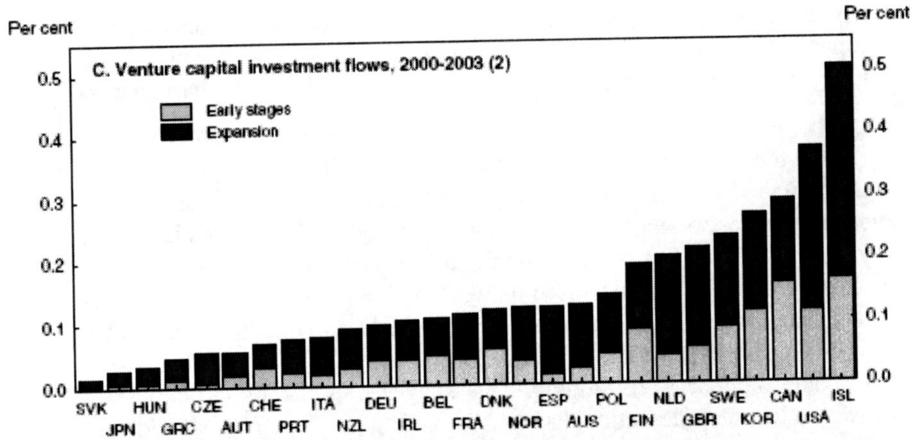

1. The asset class of venture capital or private equity (or buyout) funds dedicated to invest from funds raised by third parties into growth or restructuring cases.

2. 2000-02 for Iceland; 1998-2001 for Australia, Japan, Korea and New Zealand.

Source: European venture capital associations, World Bank Financial Development and Structure Database and OECD venture capital database.

That there is no shortage is due partly to the actions of the state. Argentum was established in 2001 with a capital base of NOK 2.45 billion, and it is a government-owned investment company that participates with minority stakes in specialised investment funds for active ownership, so-called private equity funds. A further NOK 200 million was subsequently contributed to the company. Its mission is to develop management resources for such funds and indirectly bring businesses into the private equity phase, create enhanced competitiveness in Norwegian business, and achieve a high yield on the capital invested as well as creating networks involving owners, fund managers and R&D centres. Argentum supplements the role of the national seed fund, which co-invests with the private sector at the seed, start-up and other early stage, and the more routine development banking services provided by Innovation Norway.

In total 16 seed capital funds have been authorised since 1997. Six were established from 1997 to 2000, and ten more were authorised between 2003 and 2005. The latest funds are of two types: six regional funds were authorised following budget negotiations in 2003 and 2004, and four national funds following negotiations in 2005. These funds are based on co-operation between the government and private capital, and will amount to about NOK 2.4 billion when fully capitalised. Both types offer incentives for private investors, in the form of maximum limits on the yield to the state and subsidies to cover losses on individual investments. The incentives are somewhat better for private players under the regional scheme. The funds invest in the early phases of innovative businesses with growth potential and ambition.

Policy therefore addresses a mix of national and regional policy concerns, with seed funds seen as necessary for regional autonomy. Individual, necessarily small, regional funds are preferred to a smaller number of national funds with regional distribution, something that Innovation Norway could alternatively provide easily through its network of district offices or that SIVA could offer via its extensive national network of science and industry parks, in addition to the venture funds that SIVA supports.

The tension between the need to build scale in venture funds (to build big enough portfolios for risk taking and to afford good-quality investment analysis) and the pressures of regional policy is not new. STEP's 2000 evaluation of SIVA investigated four of its funds and found that they mainly invested in rather mature companies making established products, though with a scattering of more technology-based ventures. The scale of the funds limited their ability to take risks. Ernst & Young's evaluation (1998) of the Venture and SIVA seed funds was deeply sceptical about the return of the funds, and pointed out that the state – as minority owner – was not in a position to steer their policy or development. The current and planned multiplication of small funds, many of them regional, risks renewing these problems.[44] SIVA has a very small role in the seed and venture markets, and has not entered new funds since 2004.

The problems which arise when small regional funds simply try to operate below critical mass –in terms both of analytical capacity and the absolute amount of money in the funds – have been experienced in other countries, notably Finland. In Finland the approach has been to leave management of regional fund collections to private sector managers, effectively raising the analytical critical mass.

2.7. Human resources for S&T and innovation: the flight from science

As in most OECD countries, fewer people in Norway are leaving school with good qualifications in mathematics, science and technology (MST). Over half of graduate teachers of MST are now over 50 years old. Few MSc graduates now choose a career in the education system, so the supply of qualified teachers in mathematics and natural sciences in upper secondary education seems likely to dry up in the next few years, a problem exacerbated by the rate at which people leave the teaching profession during their career because they have a lower salary and fewer opportunities for advancement than employees in trade and industry.

In recent decades, the number of students graduating with university and college degrees has increased by 23% while the percentage of degrees in MST decreased from 22.4% in 1994 to 17.8% in 2004. A dip in applications to higher education MST courses in 2004/05 has been followed by a slight recovery but the fact that MST courses are open to all raises questions about quality. 2006 admissions data show a 17.6% increase in applications to engineering, technology and architecture courses (with a rising proportion of female applicants) but a fall of 8.1% in mathematics and the natural sciences.

44. The evaluation showed that most were staffed by one or two people. When there was a larger staff, this was because the investment function was integrated into a wider range of services in incubators or development companies. While most tried to exercise active ownership, the companies in which they invested generally said that their role was limited, at best, to board representation, and that the more proactive support they needed was largely absent. The funds were essentially useless as sources of networking and business contacts. There were honourable exceptions, but these were largely due to the skills and interest of particular individuals. Few of the funds had been able to focus their investments on single industries or clusters. None was big enough to manage portfolio risk. Most were driven over time to make successively less risky investments, moving away from the seed role and towards more standard capitalism.

The number of new S&E graduates in Norway is only 76% of the EU average, but in terms of performance relative to the EU25, Norway is catching up. The percentage of women who graduated from high school in natural science, crafts and technical subjects has remained at about 25% for the last ten years.

The other Nordic countries doubled the number of doctoral degrees awarded in mathematics and the natural sciences in 2002 compared with 1990; the increase in Norway was 50% over the same period.

Changes in Norway's research specialisation can to some extent be attributed to the authorities' research priorities, such as the focus on medical research and social sciences, but may owe more to the development of the education system, whose expansion has indirectly left society's research priorities to students' choice of studies. Research activity has largely followed student figures owing to the way research-based teaching has been practised. This led in the 1990s to a growth of research in subjects popular among students and to stagnation in research in mathematics, science and technology.

The Education Ministry's initiative to address the MST issue is in part a good example of a systemic approach,including:

- A national forum for MST in work and education.
- Development of local interaction between education, trade and industry.
- Establishment of a student project for inclusion, recruitment to MST and social diversity.
- A specific focus on recruiting women.
- Increased number of teaching hours in mathematics and sciences in the school curriculum.
- A budget of NOK 600 million for competence building among teachers and kindergarten staff, where MST capabilities are a priority area.
- Development of improved methods for teaching and learning in mathematics, science and technology.
- Use of the International Polar Year (2007-08) to promote social interest in MST.

However commendable, this integrated approach to tackling the issue in the schools should be better connected to the same ministry's work on improving the quality and attractiveness of research careers, and complemented by further efforts to fill the mathematics and science gap for present, as opposed to future, generations of university students.

Chapter 3

THE ROLE OF GOVERNMENT

3.1. Introduction

Norway has emerged from being a predominantly agrarian and fishing economy by following a path of industrialisation and nation building on its raw materials base. The discovery of oil and gas was a severe shock to the established structures, but these have adapted and Norway has emerged as a productive and wealthy economy. Research and innovation policies have been important components of this development and the past two decades have seen major reforms in the agencies that support research and innovation and Norway's increasing ambitions for a research- as well as resource-based future. This is complicated, however, by a desire for increasing regional autonomy and by some governance imperfections that tend to fragment national efforts.

This chapter first briefly provides some historical perspective on the role of government in the development of the Norwegian innovation system and acknowledges the most pressing policy challenges currently facing the Norwegian government regarding the contribution of science, technology and innovation policy to sustainable economic growth and social progress. It then assesses how well existing innovation policy governance arrangements and instruments are adapted to cope with these challenges.

Despite its emphasis on path dependence, work in the innovation systems tradition often neglects to take account of history. The Norwegian innovation systems research tradition is better than most in this respect. One owes the following account partly to the Norwegian MONIT team (Remøe *et al.*, 2004), partly to the RCN evaluation and partly to other accounts of the relevant history.

3.1.1. From political independence to oil discovery

Norway was and is still a resource-based economy, with fish, wood and energy playing an important part, while agriculture has always been difficult owing to the scarcity of arable land.[45] After Norway gained independence in 1905, the idea of nation building and the country's endowment of natural resources strongly influenced Norway's economic development. Concession laws were adopted in the period from 1903 to 1917 which secured national ownership of hydroelectric power, mines, woodlands and agricultural land. Gradually the state developed these natural resources, with state-owned enterprises engaging in hydroelectric power, energy-intensive industries and oil and gas extraction, but only on a large scale after World War II (Statoil was established in 1972). Municipalities were the major developers of hydroelectric power.

Industrialisation began seriously in the first years of the 20th century, particularly with energy-intensive process industries enabled by hydroelectric power, and later with the development of a shipbuilding and mechanical engineering cluster. Many small towns were built around a single factory and a strong labour union and social democratic tradition developed. Already in the 1920s and 1930s, there was a robust labour tradition that made substantial public works a natural response to the Depression, which hit the country hard, even harder than in other parts of Europe, given Norway's reliance on shipping.[46] As in Sweden, the strength of the labour unions combined with the clear sense of "all being in the same boat" have underpinned comparatively stable labour relations and a strong partnership between the employers' federation (NHO) and the Norwegian Confederation of Trade Unions (LO), which continues to this day.

The invasion by Nazi Germany in 1940 marked a rupture in intellectual and other relations with Germany, where much of Norway's scientific community had obtained their higher degrees. Key Norwegian scientists participated in the war effort in the United Kingdom. Afterwards, the pattern of scientific links shifted to the United Kingdom and the United States, and many Norwegians obtained both first and higher degrees and established scientific networks in these countries.

45. Like Ireland, Norway experienced massive emigration in the 19th century, as a result of population growth and limited arable land. Half the population left during that century. North America was the favourite destination and there were still Norwegian-speaking villages in places like North Dakota in the early 20th century.

46. A decade of extremely high child mortality at that time profoundly affected society and is still felt today (through a very low share of pensioners in the population).

In the post-war period, Norway continued to industrialise, often in areas upstream or downstream from its natural resources: marine technologies, metals, chemicals and specialised electronics. The discovery of oil and gas in the North Sea launched a new trajectory. As discussed in Chapter 1, this led to dramatic changes in the structure of the economy. In 1970, agriculture and fishing still represented about 6% of total value added;[47] in 2006 they accounted for only 1.5% (including aquaculture, in which Norway is the world's leading producer of salmon). The share of manufacturing declined from 21% to 9.4%, while oil and gas extraction rose from 0% to 28.1%. In other terms, the contribution of offshore industrial activity increased from less than 9% in 1970 (when it consisted of shipping, prior to the start of oil production in June 1971) to nearly 31% of industrial value added in 2006. Massive growth in the oil and gas sector tends to mask changes in the contribution of other industries to GDP. If one removes the contribution of oil and gas extraction from GDP, the share of manufacturing fell from 21% to 12.6% between 1970 and 2006, while the share of services, including government services, rose. In the services sector there has been both rapid growth (*e.g.* in business services) and a pronounced relative decline (*e.g.* in wholesale and retail trade and shipping). There has been an overall shift from manufacturing towards services which, despite the efforts made, may not yet be sufficiently reflected in research and innovation policy priorities. Services are an important part of the Norwegian economy; some 75% of Norwegian employment is in the services sector, well above the OECD average.

Thus, the production structure of the Norwegian economy has some particular features;[48] government and oil-related activities account for a large part of total value added. The private non-oil sector is skewed towards small units, and there are only a few companies of international size. These big companies are more or less linked to the energy, maritime, telecommunication or financial sectors, and many still have a high share of state ownership: Statoil, Norsk-Hydro, Telenor and DnB NOR. In contrast, companies in shipping, shipbuilding and offshore supply industries are privately owned. The petroleum sector has high technology content and is very capital-intensive. It employs very few workers but generated around 30% of total value added in 2006 (productivity levels in the sector are therefore very high).

47 Measured at basic prices (Statistics Norway).

48 The remainder of this section draws on OECD (2007a).

Petroleum exports are the main source of the trade surplus (20% of GDP in 2005, crude oil, natural gas and pipelines accounting for more than 50% of export value). Excluding oil and gas, mainland exports are relatively diversified, but are mainly low-technology. The non-oil trade deficit is very large: non-oil exports of goods and services covered only 67% of imports in 2005.

Today, Norway has an economic structure that is still heavily dependent on natural resources, such as oil, fisheries, fish farming, hydro-electricity and aluminium. It has been argued that – like other resource-based economies – it may not be sufficiently diversified to protect Norway from fluctuations in global demand and world commodity prices. In addition, natural resources such as oil are depletable.

3.1.2. Research and innovation policy over time

During the post-war decades, Norwegian industrial policy focused on closing the "productivity gap" with the United States and developing large-scale industry. The main concern was to develop large manufacturing companies and to make the results of basic research available to them. Little effort was devoted to the small and medium-sized enterprise (SME) sector and, in line with the recommendations of the Vogt Committee of 1946, applied research was assigned lower priority than basic research. This policy was implemented under a series of Labour governments, enabling a compromise between the Norwegian Confederation of Trade Unions and the state, in which the unions agreed to high rates of technology deployment, rationalisation and productivity increases to allow the financing and construction of a welfare state. The state played (and still plays) a major role as entrepreneur in these developments.

From the second half of the 1970s, post-war industrial policies were dismantled, reducing the importance of state intervention, increasing technology push and introducing measures aimed at SMEs. Innovation, as opposed to research, policy was effectively introduced and became an increasing focus of regional as well as national policy. Market mechanisms began to be preferred to direct state intervention. During the 1980s, policy also shifted from promoting the development of large individual companies to more generic technologies, in the shape of nine main target areas (*hoved-innsatsområder*) in areas such as information technology (IT), oil and gas, new materials, biotechnology and aquaculture, which obtained the lion's share of the growing budget for research.

From 1978, the exploration and exploitation of Norway's newly discovered petroleum resources were used to establish "goodwill agreements" with companies that obtained concessions, requiring them to procure R&D and technological services from Norwegian suppliers. The applied research institutes benefited from this arrangement which also promoted the development of Norway's strong offshore services and supplies industry (itself a development of the existing shipbuilding and marine services sector). Also in 1978 the first SME White Paper (St. meld. nr. 7, 1977-1978) announced greater policy emphasis on developing small firms rather than continuing to reach US productivity levels by betting on large companies.

Following the report of the Thulin Committee, "Research, Technological Development and Industrial Innovation (*Forskning, teknisk, utvikling og industriell innovasjon*) (NOU, 1981), the natural science and technology research council (NTNF) had to divest itself of its applied research institutes. Their core funding was gradually reduced in an effort to ensure their relevance to industrial needs. This amounted to a rejection of the Vogt Committee's faith in the linear model. More broadly, the focus of innovation policy shifted from science to technology.

In spite of rapid increases in R&D spending through the 1980s, there was broad policy concern that Norway was still underinvesting in R&D. This was largely due to the idea that new industrial structures would be needed to underpin welfare growth. The 1989 White Paper (St. meld. nr. 28, 1988-1989) set a target of increasing R&D spending by 5% a year in real terms. Following several years of campaigning by the new Research Council of Norway (NCR), the 1999 White Paper (St. meld. nr. 39, 1998-1999) set a target of increasing the share of GDP devoted to R&D until it reached the OECD total (then a little over 2%). Subsequently, Norway adopted the Barcelona goal of 3%. Remøe *et al.* (2004) point out that the 1999 Research White Paper (St. meld. nr. 39, 1998-1999) was structured around the idea of innovation systems and marked the start of a movement within ministries and agencies towards trying to build a comprehensive research and innovation policy.[49]

Together with SIVA, which develops and runs business and science parks, RCN and Innovation Norway now form a trinity of organisations that are intended to provide a complete range of state services and funding for research and innovation. With the exception of the modestly sized Cancer Fund and the venture capital sector (which includes state-owned seed capital

49. An innovation systems approach was already by the Thulin Committee in 1981, and could, to some degree, be found in the 1983 White Paper on technological-industrial R&D (St. meld. nr. 54, 1982–1983), After disappearing for a decade it reappeared in the White Paper on Research in 1993 which argued strongly against the "linear approach".

and regional venture capital funds in addition to the private sector funds), almost the only independent source of research or innovation funding in Norway is the European Union's Framework Programme, from which Norway, like some other small advanced countries, tends to receive about as much as it contributes. The government announced its intention to top up private gifts of over NOK 5 million to universities by 25% in the 2005 White Paper on research, and the scheme was introduced as of 2006.

3.1.3. Regionalisation

For much of the post-war period, regional policy focused on redistribution of income to more remote areas in order to prevent depopulation and the creation of a vacuum in the north. Regional policy was traditionally conceived as "district policy", with the centre managing the regional distribution of resources. With local government comprising 431 local communities *(kommuner)* in 19 counties *(fylker)* managed by an appointee of the central state *(fylkesmannen,* who formally represents the monarch), there was little or no input from the regions into regional planning and development. Regional parliaments *(fylkesting)* chaired the communities until 1975, when direct elections were introduced and it became possible to develop regional policies at the level of the counties. The government has announced that it is reviewing the current county structure and the counties' administrative tasks.

In the 1980s a new regional politics culminated in the 1989 White Paper (St. meld. nr. 29, 1988-1989), which emphasised innovation and entrepreneurship in regional development. Regional colleges were created by merging smaller entities and are expected to be cornerstones of regional development. Regional innovation programmes – BU2000 and VS2010 – during the second part of the 1990s and into the 2000s used trade union participation in companies' innovation processes as a way to build social capital and improve productivity. They aimed to build on the resulting social capital to bring both the unions (through the Norwegian Confederation of Trade Unions – LO) and the employers (via the Confederation of Norwegian Enterprise – NHO) together to develop regional innovation systems. The regional universities, colleges and institutes were expected to play a strong role so as to help to support the regional economy.

Innovative activity is strongly concentrated to the Oslo area and to a lesser degree in the other main cities. There has been continuous pressure to devolve more responsibility for innovation and even research policy to the regions. In 2000, the government tasked the regions with ordering services from the SND's regional offices. More recently, further resources have been decentralised, and the Research Council has been told to assist in establishing regional research funds.

In the White Paper on Rural and Regional Policy (No. 21, 2005-2006), the Norwegian government declared its intention to give Norwegians a real choice about where they want to live and to ensure that "all parts of the country are put to use". This implies "sustaining in large measure the present settlement pattern" (Norwegian Ministry of Local Government and Regional Development, 2006). Specifically, the government seeks to develop and consolidate the educational and scientific communities in all parts of the country, and will consider how:

- Courses and student admissions can be more evenly distributed than at present.
- The funding system can ease cross-fertilisation between higher education and the business sector in various parts of the country.
- The funding system can help decentralise higher education.
- To stimulate research in the private sector and to achieve a more balanced geographical distribution of R&D.

Measures have been taken to decentralise responsibility for innovation policy from national to regional authorities, and to continue and further develop the existing incubator and industrial parks programmes. It plans to give priority to an initiative for commercialising business ideas developed at regional colleges. The proposed Centres of Expertise programme was implemented in 2006 and targets regional clusters of specialised and internationally oriented business and knowledge environments.

3.2. Main current policy priorities and challenges

Norway reached a consensus during the decline in the price of oil in the late 1980s that it does not have the option of depending in the long term on petroleum revenues. This was expressed, among others, in the Steigum Report of 1988 ("Norwegian economy in transition") and the establishment of the Oil Fund (*Statens petroleum fond*) in June 1990. Even though large amounts of state oil revenue have been ploughed into this international investment fund (now referred to as the Pension Fund), extrapolation suggested a considerable gap between the income to be expected from national industry plus the Pension Fund, and the income needed to maintain historical rates of income growth and welfare, especially since the share of old age and disability pensions in GDP is expected to rise from 9.2% in 2002 to 20% by 2050 (NHD, 2003). A recent OECD working paper also points out that, despite the Fund, Norway does not have the resources to pay for future pension needs under current rules (Bellone and Bibbee, 2006).

3.2.1. Government priorities

As a result, policy attention has increasingly turned to ways to add value to existing industry and to how the state can help to act as midwife to "unborn industry" based on research. This has been translated into priorities for S&T and innovation policies, notably in the most recent White Paper on research.

Figure 3.1. Priorities from the 2005 White Paper on Research

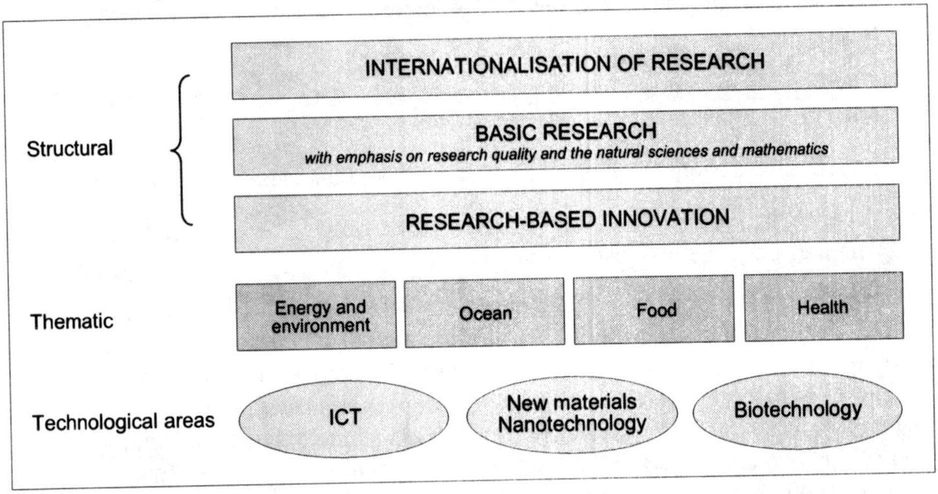

There is of course some uncertainty, since there has been a change of government since the last White Paper on Research (St. meld. nr. 20, 2004-2005) was published. However that White Paper laid down a number of priorities that have received broad support in the *Storting* and are consistent with the trend in Norwegian research policy in recent years (Figure 3.1). The priorities have repeatedly been confirmed by the current government and there has been follow-up work on most on the items cited below. These include, among others, increasing emphasis on research quality at the expense of capacity building and raising the priority of research-based innovation:

- Making Norway into a leading research nation by taking on the Barcelona goal of spending 3% of GDP on R&D, as compared with slightly more than the 1.7% reported for 2003. Two percentage points of this were to come from industry, implying an increase of NOK 6 billion in the state sector and NOK 23 billion in industry (doubling the effort, and implying, on a rough estimate, an increase of the order of 20 000 R&D personnel[50] in five years, certainly a difficult goal to reach).

- Increased internationalisation of research, involving strengthening of almost all forms of existing international co-operation worldwide, including with major powers such as the United States, Japan, China and India as well as via foreign aid.
- Increased efforts to raise the quality of research in the Norwegian knowledge infrastructure, with increased use of competition in allocating basic research resources.
- Increased budgets for instruments that aim to enhance companies' technological capabilities, including participation in EU Technology Platforms.
- More effort to involve the public in discussions of research ethics and to raise the amount and quality of science communications.
- Making research careers more attractive to younger people in order to secure the needed expansion of manpower. The White Paper promises a range of measures to increase wages, to improve conditions, to increase the number of posts such as post-doctoral positions, to encourage researchers to spend time working abroad and to help them return to Norway afterwards and to encourage gender equality in research.
- Universities and regional colleges were to be brought under the same financing model, which would include performance targets in both research and education. Special measures would be used to strengthen the research capacity of the colleges, which lags that of the universities since they have in the past largely been teaching institutions.
- The Research Council should aim to fund larger projects to reduce fragmentation of research, to strengthen the concentration induced by new instruments such as centres of excellence, and to increase investments in instruments and facilities.
- Development of a national strategy for research in mathematics, basic science and technology.
- Strengthening of the role of the research institutes via measures to support more co-operation with the universities, increasing RCN's role

50. In 2004, according to the OECD Main Science and Technology Indicators, there were 16 150 researchers in Norwegian industry and BERD accounted for 0.88% of GDP. Pro rata, this means 36 705 researchers would be needed to account for 2% of GDP, a difference of 20 555 people.

in governance by creating an indicator-driven core-financing scheme and increasing the budget available for core funding.

- RCN is to play a prominent role in monitoring progress against the goals of the White Paper.

These changes are to be implemented in a context of four research themes – energy and environment, marine, food and health – and three areas of technology – ICT; new materials and nanotechnology; and biotechnology (in effect, the research priorities of most countries since the 1980s).

These research priorities need to be embedded in a broader innovation strategy. In 2003, the last government produced a statement (NHD, 2003) on innovation policy in rather general terms, announcing that it needed to pursue a comprehensive policy in order to increase competitiveness, establish the broader economic base needed to fund Norwegian welfare, tackle the low rate of innovation in Norwegian industry, address skills shortages and raise the low national level of investment in R&D. This meant changes were needed in framework conditions for trade and industry, education, R&D and commercialisation, entrepreneurship, and infrastructure. The current government is continuing to respond to the thrust of these arguments, and, together with the announced White Paper on innovation, intends to have more detailed plans in place in 2008.

A coherent research and innovation policy requires a clear basic agreement on the contribution to be expected from research and innovation in achieving consensual societal goals, notably sustainable economic growth, and the improvements to the innovation system that are necessary to make that contribution possible. Policy discussions in Norway sometime seem to mix the two types of issues.

3.2.2. Policy challenges

The Innovation Trend Chart (Kallerud *et al.*, 2006) for Norway, written by Norwegian researchers, singles out four policy challenges: below average business investment in R&D and innovation; low public R&D funding; insufficient levels of new S&E graduates; below-average university R&D financed by industry.

These challenges may be real, but it is not clear why they should be considered more important than others and, more importantly, whether they are not symptoms of underlying problems rather than the problems themselves. In fact, the various challenges facing the Norwegian government appear to be at a range of levels. Some relate to objective changes in reality, which seem difficult to escape. Oil and gas will certainly eventually run out and Norway will have to find alternative sources of income; the diversification of production is therefore an inescapable challenge. The economy, research and innovation are all globalising, so internationalisation is another challenge to be tackled. As in many countries, young people in Norway are less inclined than before to study mathematics and science, starving the economy of these skills. Low scientific quality in research, if they should occur, would also undermine the functioning of the innovation system.

Other things are seen as challenges because they fail to conform to certain ideas about performance. The most obvious is the "Norwegian puzzle", or the fact that the Norwegian economy and innovation system appear to be in good health despite their "poor" performance against some internationally accepted indicators. A second concern is sometimes voiced about the weakness of university-industry links. This concern seems to fail to take account of the shape of Norway's knowledge infrastructure, which has a relatively larger industrial research institute sector than many other countries. In addition, while pointing to some room for improvement, such concerns should not hide the fact that university-industry co-operation as measured by innovation surveys is quite frequent by international standards (see section 2.5.1). A third is the continuing perception that Norway has a poor supply of venture capital, relative to its needs. Finally, there are recurrent discussions of research and innovation governance.[51]

Some of these challenges are discussed in Chapter 1 (the "Norwegian puzzle") and in Chapter 2 (internationalisation, the flight from mathematics and science in university enrolments, industry-university links, venture capital). Issues regarding governance are examined in section 3.4. This section discusses the challenges raised by the need to diversify production and maintain and enhance research quality.

3.2.2.1. Diversification of production

The Norwegian economy has been restructuring significantly owing to the growth of the oil and gas sectors and to the shift towards services that typifies rich countries. It should be kept in mind, of course, that a significant (but difficult to measure) proportion of the services sector supports oil and gas activities and that an important part of this activity is technical in nature.

The innovation policy response so far to the need for diversification has been broad and largely unselective. One aspect has been the creation of various business and innovation support measures, which are available to all firms through Innovation Norway (although a number of measures include regional provisions) and RCN and range from training and banking services to R&D grants and tax incentives (see section 3.4.2). This respects the

51 On the specialisation patterns and quality of research in Norway see section 3.3.2.2.

principle of neutrality, and therefore, while RCN and Innovation Norway sometimes organise programmes relevant to individual industries or clusters, it is also their principle that any and every proposal should be heard, irrespective of the sector. This principle is based not only on a concept of fairness but also on the idea that industry knows best and that an administration should not pick winners or second-guess industrial needs.

RCN's main innovation instrument is user-directed research (*brukerstyrt forskning*), in which industry defines research support needs and receives a subsidy of up to 40% (often less), which the company typically uses to finance part of external R&D, generally from a Norwegian research institute. This mechanism of channelling subsidies through industry has had a powerful effect on the direction of subsidised R&D. Both their power as customers and the high share of project costs that firms themselves bear mean that projects are fairly close to market. The institutes have had their core funding cut dramatically since user-directed R&D was introduced, so the balance of power in project definition has shifted significantly away from old-style technology push towards stronger user pull.

A second strand covers the research priorities set in recent White Papers and the generally comprehensive innovation policy that has been advertised but is not yet operationally defined. Research priorities are very wide-ranging and permissive. Given the structure of Norwegian industry and the knowledge infrastructure, there is little that is not prioritised.[52]

A third strand aims at "unborn industry" and involves a cluster of actors: SIVA, the agency for infrastructure, investment and knowledge networks, as well as innovation centres. SIVA provides support and advice to start-ups and more mature companies; the long-established FORNY programme to support commercialisation of inventions; the more recent Bayh-Dole style removal of the teacher's exception and the corresponding expansion of the industrial liaison office and commercialisation function in the universities; the extension of the state's role in innovation-related venture capital from seed funding to larger involvement in regional growth funds (while reducing the state's role in more conventional venture capital, through the sale of SND Invest, SND's former venture capital division).

52 Of course one may observe "revealed preferences". *Effective* priorities can be identified by studying the actual allocation of resources to different areas. The issue here is whether there is a clear ex ante prioritisation mechanism in place.

Implicitly, the research commercialisation strand in policy piggybacks on other applied and basic research funding, which it intends to exploit. As a result, the flow of commercialisable ideas is connected to the existing pattern of research funding. In so far as that flow already reflects national interests to some degree, this is probably useful. It is normally not the lack of new ideas but an inability to connect them to customers that is the downfall of technology push policies. Focusing mechanisms that increase the chances of new ideas arising in fields where there are customers close by are therefore helpful.

The non-selective approach – which gives due weight to a firm's own assessment of the respective projects as viable and profitable – has considerable strength in so far as industrial development most easily happens on the basis of existing resources. For example, Australia, Canada and South Africa have built successful international engineering and equipment businesses based on companies that originally supplied domestic resource-based industry, just as Norway has established internationally exploitable competences in offshore technologies.

A potential downside of this approach is lock-in to established trajectories, which may be reinforced by the use of RCN's stakeholder-based programme management boards – recognising the advantages this brings in terms of ensuring the relevance of research that RCN funds. Innovation Norway, too, is liable to a similar lock-in via the governance of its regional offices by stakeholders in the regional economy.

In innovation terms, the RCN's structure has three ingredients needed for diversifying the economy. The Innovation Division largely builds on existing industry through user-directed R&D. The Science Division is in the hands of the researcher-steered camp. The aim of the Strategic Priorities Division was to avoid the traditional opposition between user- and researcher-directed research by creating a strategic space in which new programmes could act as agents of change and the boundaries between "basic" and industrial research are dissolved. In practice, genuinely new efforts like the RENERGI clean energy programme aim to work with opportunity-creating and disruptive technologies, but it has largely become a location for some of the larger activities of the Innovation Division. RCN has begun to run foresight exercises at the level of individual technologies and industries but its ability (or freedom) to act as an agent of change appears to be underexploited. This may relate as much to the lack of a national arena for setting consensual priorities (this is discussed below as part of governance) as to reluctance to accept the risks associated with betting on some things and not on others. RCN receives substantial funding through the yield on the Fund for Research and Innovation that could be used to this end.

3.2.2.2. *Research quality*

Research quality is a concern in Norway not because there is a significant problem but because policy makers recognise its role in maintaining national competitiveness. Norway has consistently produced 0.5-0.6% of world publications in ISI-indexed journals for the last 20 years. Since the start of this century, publications have increased at a faster rate than the size of the scientific community (RCN Indicators Report, 2005). Maintaining a stable share of world production is a major achievement[53] because of the rapid growth of the world scientific community and the corresponding growth in publications.

Figure 3.2. Relative citation index, four Nordic countries, 1981-2003

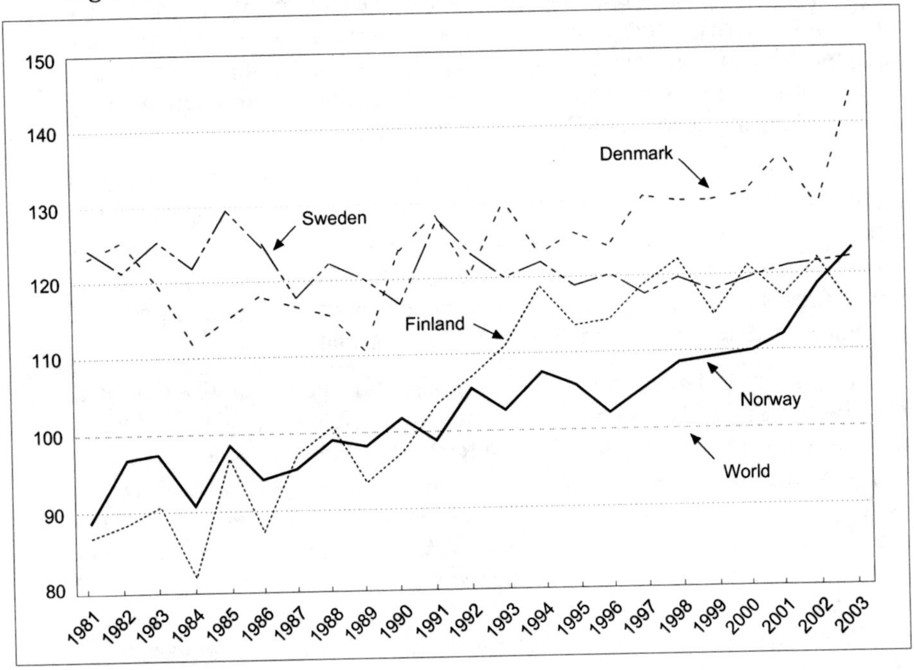

Source: RCN, 2005 (Thomson ISI data).

53. Not all bibliometric studies support this position. There is also evidence of a modest decline in world share. The differences in analysis appear to relate to differences in the coverage of the publications taken into account. For an alternative view, see Cadiou *et al.* (forthcoming 2008).

Table 3.1. Scientific publications in selected sub-disciplines

	Norway	
	Number of articles 2003-4	Percentage of world production 2003-4
Biological disciplines		
Botany	241	0.6
Zoology	235	0.9
Marine/Fish	684	2.6
Ecology	558	1.0
Microbiology	220	0.5
Agricultural science	345	0.8
Biochemistry/Medicine		
Biochemistry/Biophysics	277	0.4
Molecular biology/Genetics	265	0.4
Neurosciences	364	0.5
Immunology	233	0.7
Pharmacology	201	0.5
Physiology	108	1.1
Medicine		
Clinical medicine	3057	0.7
Dentistry	116	1.2
Psychology/Psychiatry	392	0.8
Geosciences	898	1.3
Physics/Chemistry		
Physics	650	0.2
Astrophysics	122	0.3
Chemistry	781	0.3
Technology		
Engineering	598	0.4
Materials science	145	0.2
Mathematics	195	0.5
Social Sciences		
General social sciences	523	0.8
Economics	252	0.9

Source: NIFU STEP, Thomson ISI.

In 2004 in terms of articles per thousand population, Norway (with 1.2) trailed Switzerland (1.99) and the other Nordic countries (notably Sweden, with 1.67) but was more productive than the United Kingdom (1.17), the United States (0.91), Germany (0.78) and France (0.75). Some of these differences result from different national specialisations. Small countries (such as Sweden and Switzerland) that focus on fields like the life sciences which have a high propensity to publish tend to outperform large countries with a wider range of interests.

At the aggregate level, citation data suggest that the quality of Norwegian research has been improving substantially over the past couple of decades (see Figure 3.2).

Table 3.1 indicates the fields and sub-fields in which Norwegian science is most active. The presentation in the RCN Indicators Report 2005 shows that Norwegian science specialises in the geosciences and biology (this is sometimes referred to as the bio-environmental model with geosciences as the main focus) but less so in pure sciences, physics, chemistry and engineering. The relative citation index is high in mathematics, clinical medicine, physics and geosciences.

Policy to maintain and increase scientific quality in Norwegian university research tends to follow good international practice. If the proportion of block funding is to remain as high as at present, then it would be useful to consider tying part of the block grant to research performance. Within the competitive component of funding, RCN already uses a broad mix of instruments that includes traditional response-mode ("bottom-up") funding, instruments targeted at PhD students and young researchers, and centres of excellence, to build larger research entities. RCN practice includes both traditional centres of excellence and competence centres that establish long-term research-based relationships between the knowledge infrastructure and industry, and these should over time also help to increase industrial absorptive capacity.

The main risk to scientific quality is probably the ageing of researchers, which implies that recruitment issues are critical. There may also be a risk of fragmentation. While RCN's national programmes tend to reinforce scale and therefore the strength needed to compete internationally, regional policy aims to strengthen research in the regional colleges and universities. Some level of co-ordination between the two levels would be helpful to encourage universities and colleges to specialise in areas of actual or likely comparative advantage.

Figure 3.3. Public governance of the Norwegian innovation system: institutional profile

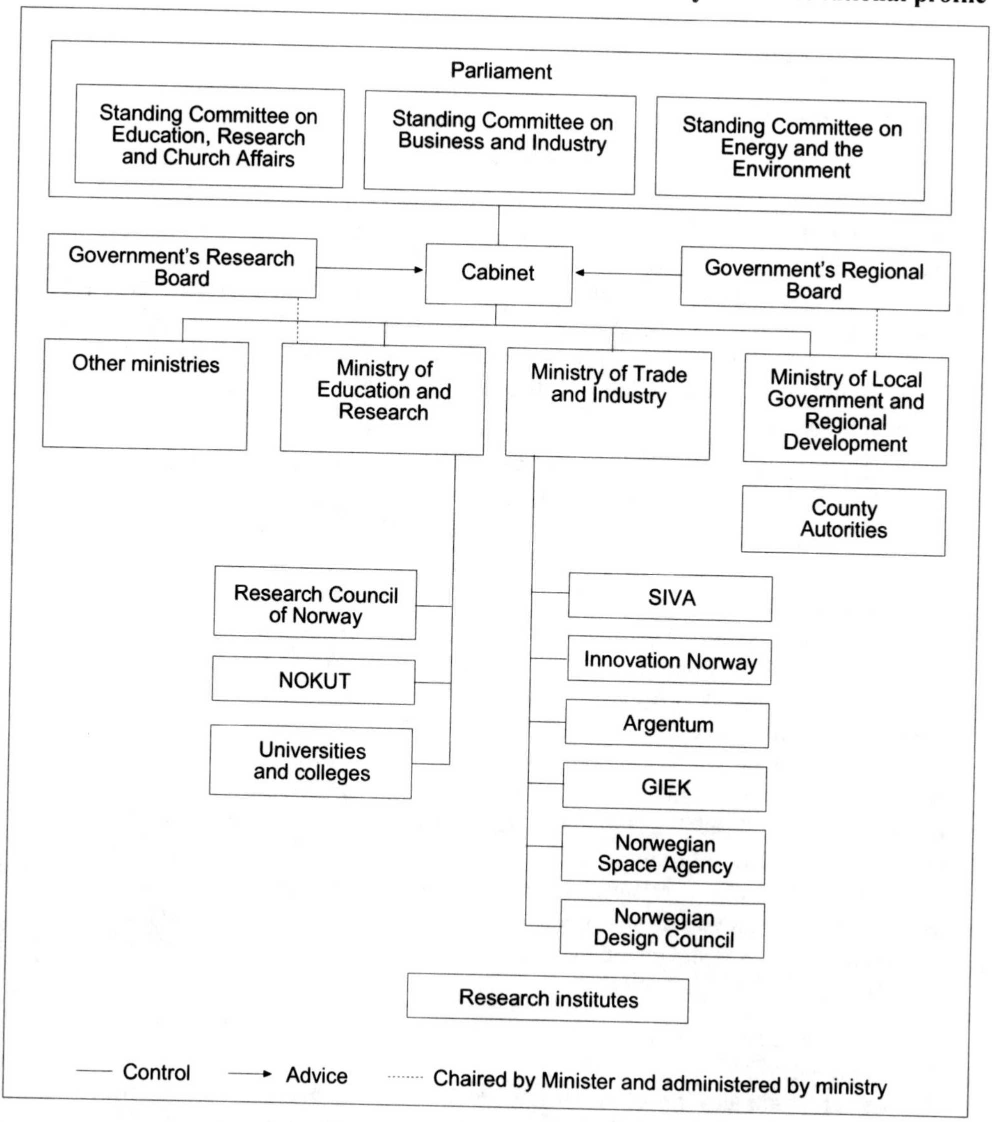

3.3. Governance and policy mix

3.3.1. Overall governance

3.3.1.1. Institutional profile

Figure 3.3 provides an overview of the components of the innovation system. The orientation of Norwegian innovation policy has been changing over the last few years. A more systemic model is gradually replacing a predominantly research-oriented linear model by integrating innovation policy into all policy areas relevant to innovation performance. This ongoing process was outlined in the 2003 document, "From Idea to Value: The Government's Plan for a Comprehensive Innovation Policy". Co-ordination of "sectoral" innovation policies is the responsibility of the Ministry of Trade and Industry.

A high-level government committee, the Cabinet Research Committee *(Regjeringens forskningsutvalg*, RFU), co-ordinates research policies within the government. The board is chaired by the Minister of Research and Higher Education and is currently made up of nine ministers and an observer from the Prime Minister's office. The formal structure and functions of this committee have remained largely unchanged from one government to the next, while its actual involvement and influence have varied over time. The Committee is typically most active in setting strategic priorities when the White Papers on research are being prepared (every 4-6 years). Meetings are held approximately every second month. The Committee receives some of its input from the Ministries' Research Committee *(Departementenes forskningsutvalg*, DFU), which consists of senior office holders in the Ministries. In addition to preparing input to the meetings of RFU, the forum discusses central issues related to R&D, identifies cross-sectoral research issues, helps promote a national research policy, and assists the Ministry of Education and Research in the administration of the Research Council of Norway.

A similar structure was established in 2004 for the co-ordination of innovation policies, the government's Innovation Board, chaired by the Minister of Trade and Industry. This board was not renewed by the second Stoltenberg government and there is currently no national arena for overall innovation policy governance. However, the government could create such an arena through its broad-based and ongoing work on the White Paper on innovation policy.

3.3.1.2. Strategic intelligence

In comparison with other countries and in addition to what is available in the international research market, Norway is well provided with national sources of strategic intelligence on the research and innovation system.

- The Research Council of Norway has historically maintained a strategic capability but it currently has few dedicated staff and is working to increase it; it also co-operates with Innovation Norway on strategic intelligence.
- The NIFU STEP institute[54], with some 70 professional staff working on diverse issues in research, education and innovation, is among the largest organisations in its field in Europe.
- Innovation Norway and a range of institutes, mostly regional, provide expert if more occasional analyses of aspects of the innovation system.
- In addition, the authorities have access to and make use of researchers in a wide range of other institutions and related fields.

All in all, Norwegian policy makers have an evidence base that is significantly better than in many countries several times Norway's size, although mismatches between supply and demand may still occur.

The use of evaluation is widespread (not least owing to legal requirements for *ex ante* evaluation and ongoing monitoring) and has actively been promoted by Innovation Norway and the Research Council both through their own practice and the national forum for evaluation, EVA. However, there appear to be two significant gaps in the strategic intelligence available to policy makers:

- While foresight (Box 3.1) is now used at the thematic and technology level, especially in relation to RCN's large programmes, Norway has yet to employ foresight at the national level to discuss overall research and innovation priorities. The lack of a national arena for overall innovation

54 NIFU STEP resulted from the recent merger of two social science research institutes in 2004. The Norwegian Institute for Studies of Research and Education (NIFU) was established in the mid-1950s a division within the Norwegian Research Council for Science and the Humanities, *inter alia* as a source of statistics on research and higher education. It was later transformed into a foundation. Studies in Technology, Innovation, and Economic Policy (STEP) started as a research group formed by the Norwegian Council for Scientific and Industrial Research. As NIFU, STEP became a foundation during the 1990s before the merger in 2003. NIFU STEP is financed by a core grant from the Research Council of Norway and also takes on assignments from the RCN and the Ministry of Education and Research.

policy governance may have acted as a constraint on the demand side. At the same time, Norway has a long-established tradition dating back to the early 1950s of regular use of long-term analysis to inform policy. During each electoral cycle, the Cabinet and Parliament discuss issues with a 30- or 50-year horizon. These discussions are based on available information in long-term programmes and outlooks that present and aim to integrate all policy areas. Foresight studies may contribute to and gain from this practice.

- SND and SIVA were evaluated in 2000 and the RCN in 2001, so the information on the performance of the three key agencies in the research and innovation system (see below) is becoming out of date.

Box 3.1. Foresight studies

While the strategic planning tradition evolving in the 1960s made use of forecasting and scenario techniques, the term 'foresight' entered policy use following the publication of Ben Martin and John Irvine's book *Research Foresight: Priority-Setting in Science* in 1989. The 'classical' definition of foresight by Ben Martin is "the systematic attempt to look into the longer-term future of science, technology, the economy and society, with the aim of identifying the areas of strategic research and the emerging of generic technologies likely to yield the greatest economic and social benefits" (Martin, 1995). Martin stresses that foresight is not the same as forecasting – it is a process that aims to *construct* the future by integrating science and technology push with social pull, rather than a technique that tries to predict what the future will look like. It therefore uses a range of techniques to involve experts and stakeholders in developing consensus about desirable directions of change. In recent years, Martin has begun to argue that foresight is a way of "wiring up the national innovation system" because of its emphasis on building links among actors, and that this helps build systemic policies and mechanisms to strengthen innovation systems (Martin and Johnston, 1999).

3.3.1.3. Issues in governance

Governance of the research and innovation system is a traditional subject of debate in Norway. Repeated and only partially successful attempts have been made in recent decades to build a national arena or advisory functions to set directions and co-ordinate policies. One reason for this is probably that the "sectoral principle" – the idea that a ministry has sole responsibility for its area of specialisation – is deeply rooted in the Norwegian administrative and political culture. In practice, as in many other countries, overall co-ordination of sector ministries is left to the Ministry of Finance, and in their respective fields the Ministry of Trade and Industry and the Ministry of Education and Research.

Figure 3.4. Generic organisational structure for research and innovation policy

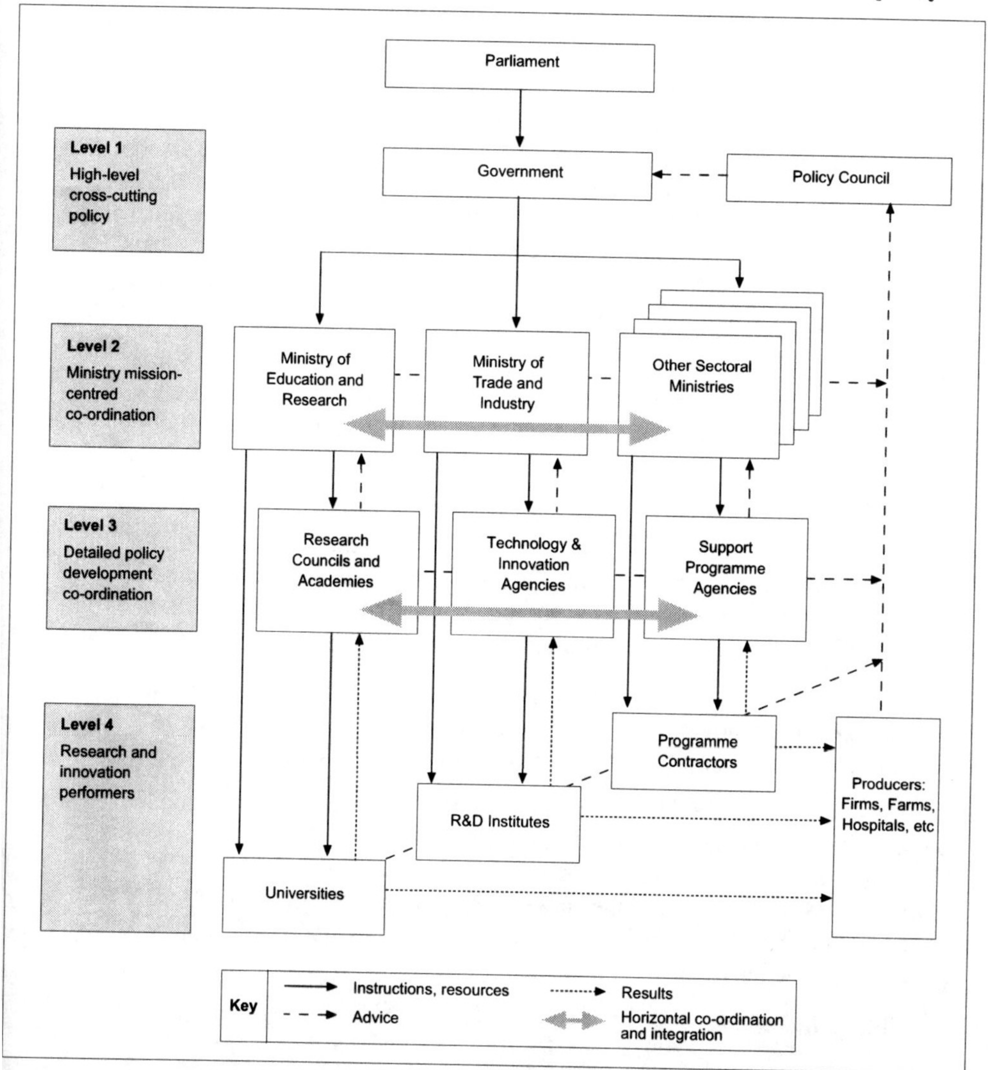

Source: Developed by Erik Arnold, Technopolis, in collaboration with Martin Bell, SPRU, in a project for the National Science and Technology Development Agency of Thailand in 2002

Studies of research and innovation governance suggest that there is no single "optimal" pattern of research and innovation governance. Figure 3.4 shows a simple model[55] of research and innovation organisation and governance which does not represent any particular national practice. This scheme has four levels of policy co-ordination:

- *Level 1* involves setting overall directions and priorities across the whole national innovation system. In many countries, this involves high-level councils (and/or high-status institutions such as academies of science and engineering) giving advice to government. A number of OECD countries have implemented more binding means at the cabinet level. These efforts are typically designed to solve a similar set of problems but the concrete approaches adopted vary considerably, depending on the historical set-up and style of government.
- *Level 2* is co-ordination among ministries whose sectoral responsibilities otherwise encourage them to pursue independent policies. In practice, this level of co-ordination may involve administrative aspects, policy issues or both. Sometimes an inter-ministerial group also functions as the Level 1 co-ordination mechanism.
- *Level 3* is more operational and attempts to draw the actions of funding agencies into a coherent whole. This level, too, can involve administrative co-ordination as well as more substantive co-ordination of funding activities, such as co-programming.
- *Level 4* involves co-ordination among those who actually perform research and innovation. Co-ordination at this level tends to be achieved through self-organisation rather than formal mechanisms.

Most of the vertical flows shown are formal. They amount to or concern *de facto* "performance contracts" between institutions at the different levels. The exception tends to be flows into the policy council, which tend to be people-based rather than paper-based, and therefore informal. Surveys (Arnold and Boekholt, 2003) suggest that key research and innovation governance functions include:

- Setting directions.
- A referee.
- Horizontal co-ordination.

55. This was developed by Erik Arnold, Technopolis, in collaboration with Martin Bell, SPRU, in a project for the National Science and Technology Development Agency of Thailand in 2002.

- Co-ordinating production among knowledge producers.
- Intelligence.
- Vertical steering.
- Enhancing the profile of research and innovation.

There has been some debate about how well Level 1 has worked in Norway. The RCN evaluation (Arnold *et al.*, 2001), for example, addressed this question and suggested that the Finnish model could work well in Norway and had the potential to resolve some of the policy co-ordination issues.

Government policies and budget priorities are set out in the annual budget process and in less frequent White Papers. An instrument for research policy co-ordination is the Government's Research Board (*Regjeringens forskningsutvalg*) bringing together research-relevant ministers under the chair of the Minister for Education and Research. For a while in the early 2000s, the Ministry of Trade and Industry operated an equivalent innovation board, which the current government abolished as a contribution to administrative simplification, leaving Norway without a secondary discussion arena at the government level.

The ministerial-level research board is mirrored at Level 2 by a co-ordination group of senior civil servants *(Departementenes forskningsutvalg)*. When called for, the Ministry of Education and Research also initiates and co-ordinates processes among relevant ministries on specific areas. Recent examples are the new funding system for the research institutes, the international polar year, research initiatives in the government's High North strategy, and the climate research initiative.[56] The Ministry of Education and Research is responsible for co-ordinating research policy in the annual budget process. This implies ensuring that priority in the various ministries is given to research initiatives that are in line with the overarching national priorities. The return on the Fund for Research and Innovation is channelled through the Ministry of Education and Research's budget. This funding mechanism has emerged as a useful tool for co-ordination and has enabled the government to initiate a number of research initiatives that cut across sectors.

56. The Ministry of Education and Research also has overall responsibility for Norwegian participation in international research collaboration, including the co-ordination of participation in the EU's Framework Programmes for research. The Ministry also chairs an intergovernmental committee that coordinates Norwegian policy towards European research collaboration.

In principle, at Level 3 the unique structure of research funding in Norway -- with a single agency spanning academic research and industrial R&D, and with Innovation Norway spanning issues of business development and innovation -- provides major opportunities. These include the potential to adopt a cross-disciplinary approach to research policy, to link academic and industrially oriented research, to connect innovation support to business development measures, and to identify and exploit opportunities that in other countries cut across the areas of responsibility of different ministries and agencies. The background of reforms in public sector administration is similarly promising. Norway has enthusiastically embraced many ideas of the New Public Management, including the principle of management by objectives. Rather than try to micromanage the work of colleagues and agencies, policy makers should set broad objectives and monitor progress using performance indicators.

In practice, having two multi-principal agencies at the centre of policy implementation has not resolved all co-ordination problems. In line with its mandate, RCN has provided the government with advice on research and innovation policy since it was set up, but it has not always been in a position to adopt a comprehensive and broadly accepted approach to the research and innovation system. Innovation Norway is closely steered by its four funding Ministries and by the fact that each of its offices has a regionally based governance structure. Neither of these agencies has a mandate to co-ordinate policy, however. Organisations like RCN "integrate" rather than co-ordinate policy (Remøe *et al.*, 2004). Both the RCN evaluation in 2001 (Arnold *et al.,* 2001) and the MONIT study (Remøe *et al.*, 2004) found research policy co-ordination to be weak.

An additional co-ordination mechanism is the regular production of a White Paper by the Ministry for research purposes and which may constitute a good arena for innovation policy discussion. The MONIT study's survey found that only a minority of stakeholders thought the innovation policy effort without such a White Paper at the beginning of the decade had improved horizontal co-ordination. However, this effort from the Ministry of Trade and Industry rather paved the way for the first Innovation White Paper, to be presented during 2008. The authorities should use the opportunity to resolve co-ordination issues in innovation policy setting in this Paper.

There is consensus that each ministry should have a research strategy, which requires considering what research the ministry needs in order to pursue its sectoral mission. A further part of the responsibility (not always stated) is to ensure research capacity in the Norwegian system (or reliably accessible elsewhere) to deliver the research the ministries need. The education and industry ministries both have responsibility for overall co-ordination but are in a different position; they also need to be patrons of research done in and for their constituencies. Integrating these different needs into commonly managed programmes is a task that RCN can do fairly easily.

A coherent innovation policy, however, requires a stronger inter-penetration of ministry roles than is the case today. For example, the industry and environment ministries would need to work together to create a comprehensive policy for encouraging environmental innovation that would build the Norwegian environmental industry and create a virtuous circle of environmental improvement and industrial success. This might include elements as diverse as technology, subsidy, regulation, trade policy, FDI, fundamental research, measurement and inspection, and would be beyond what an agency at the level of Innovation Norway or RCN could tackle on its own.

Recent changes in regional policy have increased problems of policy co-ordination. Until the early 21st century, county development plans were discussed with the ministry for regional development. Therefore, there was a certain amount of co-ordination with that ministry and – since it funded much of the regional innovation activity of SND/Innovation Norway – also with that regional agency, if not with the industry and education ministries. The decentralisation of many innovation funds – with no strings attached – to the regional level is unusual in Europe in terms of the degree of discon-nection it implies between the national and regional planning levels (OECD, 2007b, p. 81) and complicates co-ordination. In the past, there was no formal co-ordination mechanism for the ministries involved in regional development. In 2005, a government sub-committee on rural and regional policy was established. It is chaired by the minister for local government and regional development and brings together the ministers for government administration and reform; agriculture; fisheries and coastal affairs; culture and church affairs; trade and industry; and transport and communications.

There are probably positive aspects to the absence of strong hierarchical co-ordination of research and innovation policy. These would certainly include the fact that many conflicting interests get "bottom-up" access to policy attention and support. This also means a tendency to lock-in to the *status quo*. The key question is whether this should be counterbalanced by forces for change, with some more adventurous attempts to try to tackle issues such as the need to diversify the industrial structure and to deal with broader socio-economic questions. These big questions appear hard to tackle in a fragmented governance system driven by the sector principle. There thus appear to be strong reasons to reinforce the institutional mechanisms for discussing and strategising about research and innovation policy and setting broad directions.

Figure 3.5. Public funding of R&D performed in the public and private sectors

2004 or latest year available

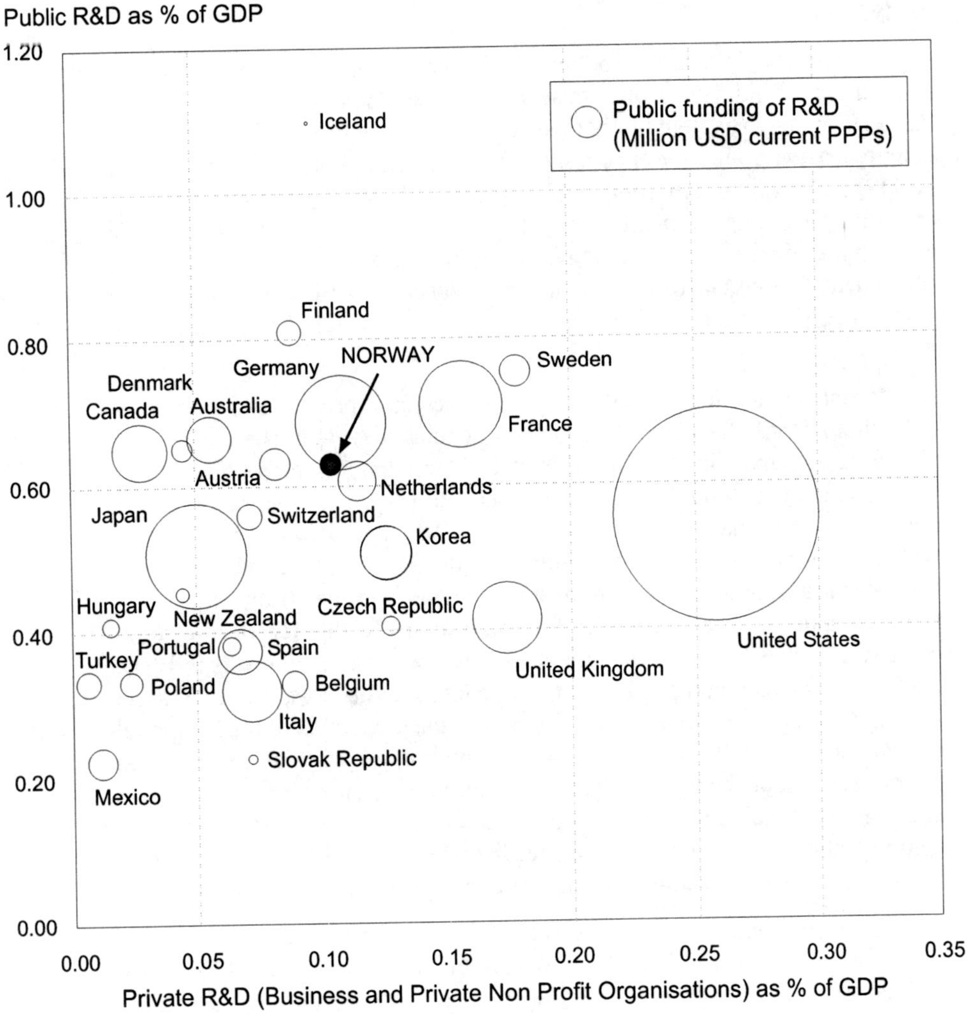

3.3.2. Policy mix

3.3.2.1. Overall balance

Through the combined action of the RCN, Innovation Norway and SIVA (see section 3.2.2 below), Norwegian innovation policy appears to perform in a rather balanced way the three following main strategic tasks: ensure an optimal rate of public and private investment in human resources for science and technology (HRST), R&D and innovation; enhance the innovation competencies of firms, especially SMEs; and foster productive relationships between the different actors of the innovation system.

Regarding public support to R&D (Figure 3.5), Norway relied for a long time on direct funding in the framework of user-directed programmes. A fundamental change occurred in 2002 with the introduction of the indirect support scheme Skattefunn (see section 3.4.1 below). Among smaller OECD countries, this approach was consistent with practices in Australia, Ireland and the Netherlands but differed from that of Belgium, Finland and Iceland (Figure 3.6). From 2002 to 2005 this resulted in substantial cuts in appropriations for user-directed programmes. But from 2006, direct R&D grants to industry have increased strongly (from NOK 599 million in 2005 to an estimated NOK 1.1 billion in 2007) while expenditure through R&D tax credits decreased from NOK 1.2 billion in 2005 to an estimated NOK 900 million in 2007 (for an international comparison, see Figure 3.7).

Two other dimensions of the policy mix concern the balance between basic and more applied research on the one hand and the alignment of research orientations with Norway's strategic needs on the other. Compared to other OECD countries, at 0.28% of GDP (2005) Norway does not seem to underinvest in basic research, given its below-average overall R&D intensity. The issue is rather to ensure high-quality, non-targeted long-term research (see section 3.2.2.2 above). As regards the second dimension there appears to be room for improvement mainly by further strengthening the governance of the innovation system, including better use of the comparatively rich Norwegian sources of strategic intelligence on research and innovation policy (see 3.3.1 above).

Figure 3.6. Public financial support to firms' R&D, by instrument

2004 or latest available year

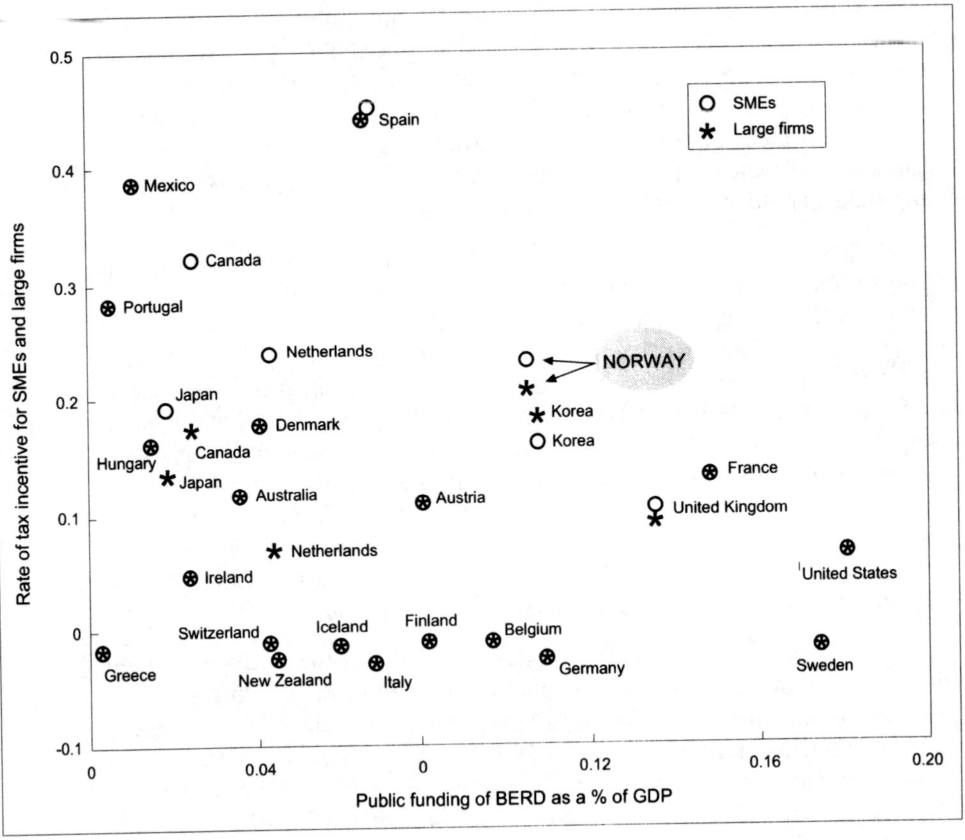

Figure 3.7. Estimated revenue losses due to R&D tax incentives as a % of GBAORD, 2005

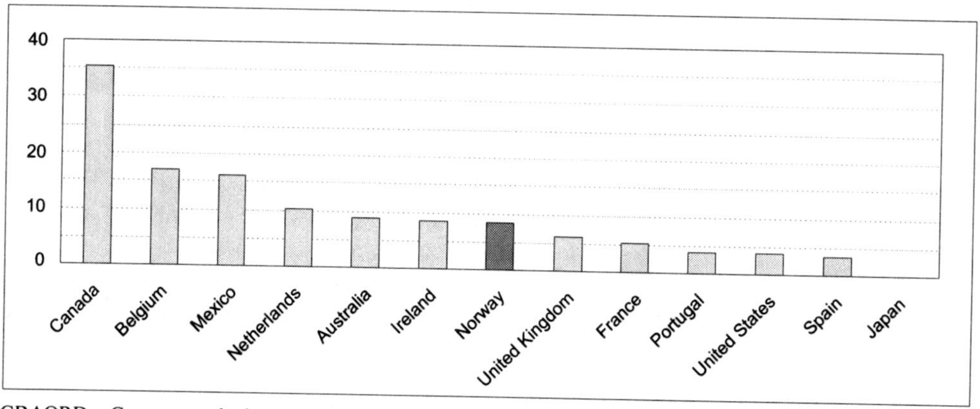

GBAORD = Government budget appropriations or outlays for R&D.

Figure 3.8. R&D expenditure by source of financing and by performing sector, 2005

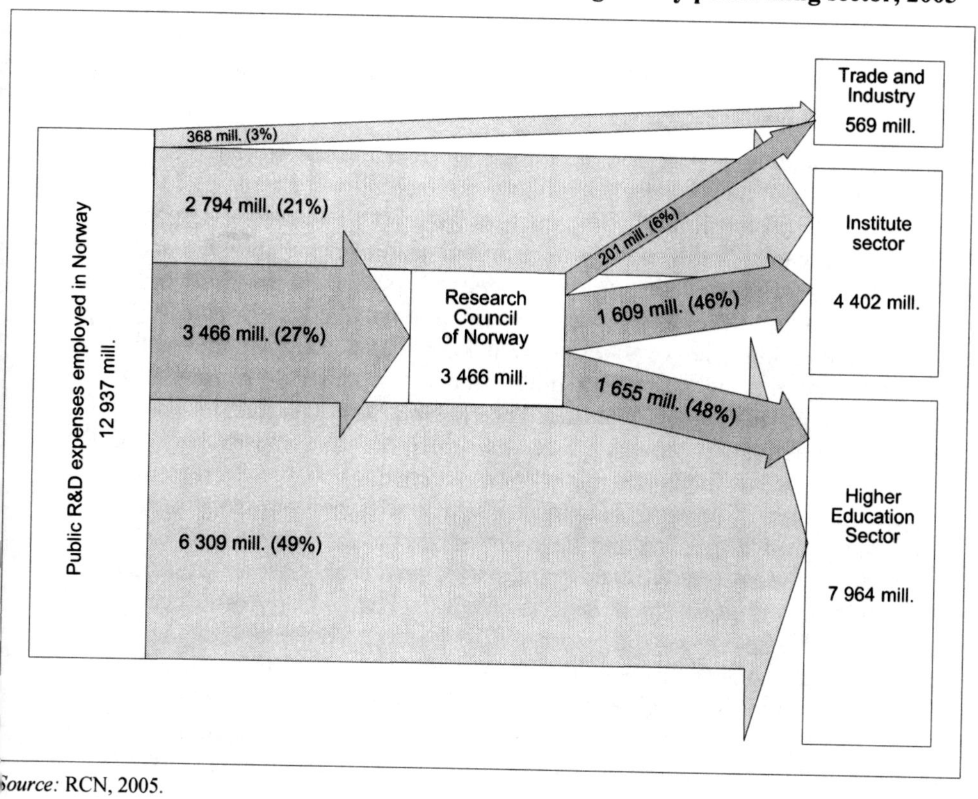

Source: RCN, 2005.

3.3.2.2. Allocation of public R&D funding

Figure 3.8 shows how public R&D expenditure is channelled to main sectors of performance of research. The picture is dominated by the university block grant. NOK 1.7 billion was channelled to the universities through RCN, so that only 21% of higher education R&D appropriations are currently quality controlled via project-level competition, a low figure by international standards. However, the recent (2006) decision to award part of the research component of university block grants on the basis of quality indicators increases the degree of quality control in the system. In contrast, RCN supplies and controls 37% of state expenditure on institutes, the remainder coming primarily in the form of block grants from ministries to institutes that support their sectoral missions. Overall, 62% of state R&D appropriations go to higher education, 34% to the institutes and 4% to industry. This compares with 59%, 36% and 5%, respectively, in 1999.

In 2005, 48% of RCN's budget went to the universities and 46% to the institutes. Only 4% (NOK 201 million) went to industry, which also benefits from the Skattefunn R&D tax incentive. RCN's expenditures have shifted significantly in favour of the higher education sector. In 1999, 57% of its spending went to the institutes and 36% to higher education, while industry received 7%. However, direct R&D grants to industry have increased strongly since 2005.

Public funding of research in Norway is based on the so-called sector principle. Each ministry assumes primary responsibility for research in and for its area of responsibility. The ministries have both a long-term responsibility for research in the sector – their broad sector responsibility – and a responsibility for research that meets the ministry's own needs for a knowledge base for policy development and administration. Sector responsibility implies maintaining an overview of the knowledge needs of the sector, securing funding and promoting international research co-operation. The sector principle is intended to ensure that research forms an integral part of policy formulation. At the same time the principle – and consequently the division of research funds into many grants – generates challenges regarding the co-ordination of research assignments, which often concern many sectors. The Ministry of Education and Research is by far the largest spender on research via the university block grant, and is responsible for co-ordinating research policy at government level (Table 3.2). The Ministry of Education and Research also has general responsibility for basic research through allocations to the higher education institutions and to the Research Council of Norway. The turn of the millennium saw a significant innovation in the way public funding of R&D is secured. As Table 3.2 shows, the appropriations from the Ministry of Education and Research have increased strongly since

2002. This is to a great extent a result of increased research policy co-ordination through the return on the Fund for Research and Innovation.

In 2008, the return from the fund amounts to over 17% of total public appropriations for R&D. The return from the Fund for Research and Innovation is channelled through the Ministry of Education and Research's budget. This funding mechanism has enabled the government to initiate a number of research initiatives. Returns from the fund are used to supplement ordinary ministerial budget allocations for research and are targeted toward research policy priorities, including quality-promoting measures, internationalisation efforts, sectoral and cross-sectoral initiatives, long-term basic research in general and long-term research in the structural, technological and thematic priority areas stipulated by the government. The fund was originally designed to ensure more stable, long-term public financing of Norwegian long-term basic research. As the fund has grown (it was initially NOK 3 billion), its scope has widened. The government's proposed R&D budget for 2008 will bring the fund up to NOK 66 billion.[57]

R&D performance is extremely concentrated in the Oslo capital region. The second-largest regional performer is Sør-Trøndelag, home to the Norwegian University of Technology (NTNU) and the major applied research institute, SINTEF. Hordaland has the University of Bergen and a number of research institutes while Rogaland is centred in Stavanger, which has strong institutes and a new university. The University of Tromsø makes Troms in northern Norway a significant research performer.

57 A substantial portion of the yield from the increase has been earmarked to cover Norway's participation fees for the EU Seventh Framework Programme from 2008.

Table 3.2. Estimated R&D appropriations 2002-07, by ministry

NOK millions

	2002	2003	2004	2005	2006	2007
Ministry of Foreign Affairs	378	401	406	451	481	535
Ministry of Education and Research	5 531	6 281	7 730	7 494	8 177	8 517
Ministry of Culture and Church Affairs	71	76	76	87	96	104
Ministry of Justice and the Police	19	25	29	26	26	28
Ministry of Local Government and Regional Development	71	163	154	158	155	166
Ministry of Labour and Social Inclusion	63	80	86	163	161	175
Ministry of Health and Care Systems	739	799	862	1 417	1 560	1 656
Ministry of Children and Equality	39	49	54	56	68	72
Ministry of Trade and Industry	1 685	1 472	1 101	1 175	1 333	1 505
Ministry of Fisheries and Coastal Affairs	794	620	588	617	660	705
Ministry of Agriculture and Food	429	454	453	450	490	520
Ministry of Transport and Communications	176	198	194	200	222	234
Ministry of the Environment	392	378	378	391	432	451
Ministry of Government Administration and Reform	456	342	178	35	92	239
Ministry of Finance	64	72	71	77	79	85
Ministry of Defence	823	850	865	880	895	885
Ministry of Petroleum and Energy	323	310	342	433	531	503
State banks	101	22	169	173	265	267
Total	**12 154**	**12 592**	**13 737**	**14 283**	**15 723**	**16 646**

Source: NIFU STEP.

3.4. Portfolio of instruments

3.4.1. R&D tax incentives: the Skattefunn

In 2002, Norway introduced the R&D tax incentive (*Skattefunn*) described in Box 3.2. Despite the hurdles facing applicants (not all firms are aware that they can benefit even if they do not pay much, or any, tax, and many firms' tax accountants are unhappy with the extra burden of responsibility), the Skattefunn proved very popular upon its introduction. It quickly became the most important individual source of fiscal support to private-sector R&D (some Research Council grants were phased out when the Skattefunn was introduced). Nevertheless, the amounts finally paid out each year have been consistently below the amounts initially budgeted (*i.e.* the estimated *ex ante* value of tax rebates for qualifying projects), and by a widening margin (82% in 2003, 73% in 2005), possibly because enterprises had difficulty in finalising all approved projects within the annual time limits. Furthermore, the *ex ante* amounts requested have also been falling, certainly in real terms. On average, 1.3% of all Norwegian enterprises (except public sector and agricultural) had Skattefunn projects under way in 2005, and the total tax expenditure amounted to about 0.1% of GDP.

Box 3.2. The Norwegian *Skattefunn*

The Skattefunn is a tax credit scheme for supporting innovation in Norwegian enterprises. It was introduced in 2002, and originally applied only to SMEs. In 2003 it was extended to all enterprises subject to taxation in Norway. Its main features are:

- Claims for the Skattefunn tax rebate can only be made for projects approved by the RCN; approval by the RCN applies only to R&D content, and must be achieved before end of the first year for which a tax credit will be claimed.
- Enterprises must submit their proposed projects online to the Skattefunn Secretariat in the Research Council of Norway by 1 September to have a guarantee that the application will be processed in the current year. Applications are evaluated by project officers in regional offices of Innovation Norway, who make recommendations to the Skattefunn Secretariat, whose project officers make the final decision. However, the RCN's decisions can be appealed to a special complaints body that considers only Skattefunn applications (three members). Legally the scheme operates pursuant to tax law under the Ministry of Finance, and the tax authorities may control the tax credit claims as part of the enterprise's tax statement. However, tax authorities can only verify whether costs are correctly specified and belong to the approved project and that total subsidies do not infringe state aid regulations. They are not authorised to question the R&D content.
- There are no regional or sector constraints or any constraint on conducting projects jointly with R&D institutions abroad (after certification by the RCN).
- There is no ceiling on the Skattefunn tax expenditure budget at the national level.
- Applications must describe:
    - The main goal, specific and demonstrable, of the project, together with any individual intermediate sub-goals.
    - A summary of the project describing its objectives, how the results will improve or develop new products, services or processes, and how the project will be carried out, including which partners and R&D institutes will be involved.
    - Expected milestones (prototypes, reports etc.).
    - Expected costs, broken down into salaries and overheads, purchase of R&D services, purchase of equipment, and other operating costs.

.../...

Box 3.2. The Norwegian *Skattefunn (continued)*

  - Financing, broken down into own funding, EU funding, other public funding, other private funding, and the estimated Skattefunn rebate.
  - R&D challenges, relative to the competences of the applicant.
  - How acquired competences (*e.g.* from an R&D institute) will be transferred to the applicant enterprise.
  - How the project will raise the level of knowledge within the applicant enterprise.
  - The R&D experience in the applicant enterprise: number of R&D projects in the past three years, including with co-operating R&D institutes, how R&D activities are organised within the applicant enterprise, and the numbers and competence levels of their R&D staff.
- Two levels of tax rebates are available, depending on the size of the applicant enterprise:
  - A 20% rebate for firms employing fewer than 250 persons and with an annual turnover not exceeding EUR 40 million, or balance sheet total not exceeding EUR 27 million, and not more than 25% owned by a large enterprise.
  - An 18% rebate for large enterprises.
- There is no limit on the number of projects for which an enterprise can claim the rebate, but the maximum tax rebate per enterprise is NOK 4 million for R&D projects conducted entirely within the enterprise, and NOK 8 million for projects conducted jointly with an R&D institute.
- If the rebate exceeds the assessed taxes (income tax, wealth tax, including social security contribution) of the enterprise, the difference (which can be as high as 100%) is paid to the enterprise as a grant (about three-quarters of the total tax expenditure under the Skattefunn was distributed as grants in 2005).

Initially, the Skattefunn was confined to SMEs, perhaps on the grounds that SMEs need more state aid because they demonstrably perform less R&D (adjusted for turnover, employment, sectoral effects etc.) than large firms. Although the scheme was soon extended to large enterprises, their maximum benefit is lower (restricted by EU/EEA regulations on state aid), and its maximum absolute size (less than EUR 1 million even for a co-operative R&D venture) must be considerably lower relative to turnover and average R&D spending for larger firms. This does not mean that there is no incentive effect at the margin, but if there is an incentive effect, it would be larger if the ceiling were higher.[58]

58. The R&D tax credit is tax exempt, so that 18% equals the maximum pre-tax subsidy of 25% for experimental development in large enterprises.

It is possible, but difficult to prove, that applications from manufacturing enterprises are more likely to be accepted than applications from service enterprises. This is because of the inherently greater difficulty in defining precisely the goals, challenges, and perhaps even the distinction between research staff and high-level line staff. According to Statistics Norway data, the number of manufacturing firms that applied for Skattefunn credits in 2006 was about the same as those in the private services sector, and together, these two sectors accounted for about 80% of the total. It is striking that the average size of the projects for which Skattefunn tax credits are requested is only about NOK 2 million (less than a quarter of a million euros), well below the maximum allowable.

All tax expenditure schemes have disadvantages as well as advantages, and the Skattefunn is no exception. It is worthwhile essentially if the ongoing value of the extra innovation created by the scheme exceeds the costs of the distortions elsewhere in the economy created by the higher taxes needed to compensate the Skattefunn tax expenditures. Estimating the net effect of the Skattefunn is a challenge, but Statistics Norway has recently published a detailed evaluation of the Skattefunn scheme, based on an advanced statistical and econometric analysis. The main conclusion is that for SMEs, the Skattefunn does stimulate additional R&D spending, and to an extent significantly greater than other fiscal incentives, such as grants and subsidies. The best estimate (although subject to wide error margins) is that every krone of Skattefunn tax credit results in R&D spending of 2-3 krone, compared with between 0.5 and 1.5 for alternative fiscal incentives.

However, the structure of the Skattefunn, especially as initially conceived, left something to be desired, as recognised by the authorities, who have taken some measures to combat distortions. It contains inherent rent-seeking incentives, given that the size of the tax rebate (or grant in lieu) is based on firms' reported costs for specific projects approved *ex ante* (*i.e.* through the first project year), which may be difficult to verify *ex post*. In order to qualify for the maximum possible tax rebate, it is easier for firms to inflate their cost estimates on a small number of projects, rather than present correct cost estimates for a larger number of projects. The deadweight tax loss to the economy is higher, and the potential benefits from successful innovation are lower. Investigations by the authorities showed that some firms were claiming very large numbers of hours worked by R&D personnel, as well as wage rates for them which also seemed unusually high. In addition, the standard procedure applied by the Research Council for calculating personnel costs, including overheads – nearly triple the wage level – gave firms further incentives to inflate their wage cost estimates.[59]

59. This is the same procedure as used by the Research Council for costing grant requests.

Some indirect evidence that Norwegian firms may have been operating in this way comes from Eurostat data on the reported wage share in business-sector R&D expenditure, which puts Norway at the top, at about 65%. This is the case even though Norwegian wage levels for researchers are not particularly high by international standards. A report by the European Industrial Research Management Association (EIRMA)[60] puts the average annual salary for Norwegian researchers at about EUR 60 000, a figure which is broadly comparable with researcher salaries in Austria, Denmark, Germany, Ireland, Sweden and the United Kingdom, but well below those in Switzerland and Japan. The wage share is less than 50% in the United States, where wage rates for researchers are about the same as in Norway in dollar terms, while the reported wage share in Switzerland is under 60%. When adjusted for living costs, the EIRMA data puts Norwegian researcher salaries well below those in most comparable industrialised countries.[61]

In response to the possibility that some firms were taking unfair advantage of the particular design of the Skattefunn, the rules were changed as from the 2007 fiscal year. The maximum hourly rate that can be claimed for R&D personnel has been set at NOK 500 and the maximum annual number of hours worked that can be claimed has been set at 1 850. This combination gives a maximum annual personnel cost of NOK 925 000 compared with a Norwegian average wage for all employees (full-time equivalent) of under NOK 400 000. The Ministry of Finance estimated that this modification – which was bitterly opposed by industry – together with improved administrative procedures in the RCN and the tax authorities, would save about 12% of the total tax expenditure. It is possible also that firms will have a greater incentive to embark on more projects.

Nevertheless, the Skattefunn has several desirable features in theory, especially its neutrality in accepting proposals for review irrespective of their sector, region and the tax liability position of applying enterprises. But given that projects have to be approved *ex ante* (*i.e.* through the first project year), and by staff who work in the Research Council, which derives its finance from several ministries, there is a possibility that consciously or not, the selection of projects is influenced by regional and sectoral biases. At least, there may be a tendency to favour firms with a good track record of innovation – or at the other extreme to be overly generous with firms that have yet to prove their commercial viability. The fact that 75% of the tax

60. www.eirma.org/f3/local_links.php?action=jump&doi=eiq-2007-011-0002.

61. Another particularity of Norwegian researcher salaries is their relatively flat structure across working lives. Experienced researchers in most countries typically earn two to three times as much as those embarking on their careers. In Norway, the differential is about 50% (www.cpu.fr/telecharger/Bruxelles/PCNM-salaires-chercheurs-europe.pdf).

credits are disbursed in effect as grants to firms that have no taxable income supports this reading.[62]

In international comparisons, fiscal support for innovation in Norway is not exceptionally generous (Table 3.3 and Figure 3.9). The OECD's "B-index" measures the generosity of tax incentives to invest in R&D on the basis of the pre-tax income necessary to cover the initial costs of one dollar of R&D spending and pay corporate taxes on one dollar of profit. The B-index measures only the tax subsidy element for qualifying firms and projects, and may overstate generosity in cases, *e.g.* Norway, where the ceiling on qualifying projects is comparatively low.

Table 3.3. R&D tax incentives in OECD countries, 2005

	Large firms		Special treatment for SMEs	
	Tax credit	**Tax allowance**	**Tax credit**	**Tax allowance**
Volume	Canada (20%) Japan (8-10%) Mexico (30%) Netherlands (14%) **Norway** (18%)	Belgium (113.5%) Czech Republic (200%) Denmark (150%) **Poland** (130%)[1] **United Kingdom** (125%)	Canada (25%) Italy (30%) Japan (15%) Netherlands (42%) **Norway (20%)**	Belgium (118%) **Poland** (150%)* **United Kingdom** (150%)
Combination (volume/incremental)	France (5%-45%) Korea (7%-40%) Portugal (20%-50%) Spain (30%-50%)[1]	Australia (125%-175%) Austria (125%-135%) Hungary (100%-300%)	Korea (15%-50%)	
Incremental	Ireland (20%) United States (20%)			
None	Finland Iceland Switzerland Turkey	Germany Luxembourg Slovak Republic Greece	New Zealand (under consideration) Sweden	

Bold indicates incentive introduced after 2000.

1. Only for enterprises that obtain at least 50% of their income from the sale of their R&D results.

62. The comparable UK scheme also shares this feature: most of the disbursements are grants to firms that have no or insufficient taxable income to qualify for the tax rebate.

Figure 3.9. Generosity of fiscal support to R&D in OECD countries

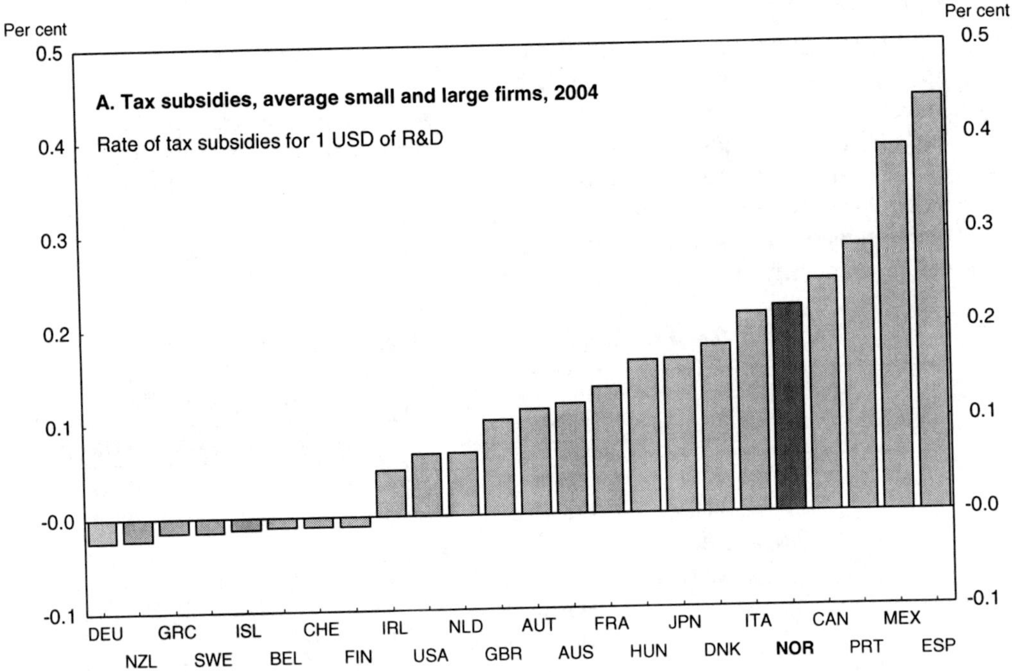

3.4.2. Public institutions and programmes for the promotion of R&D and innovation

The three Norwegian agencies that provide a comprehensive set of instruments and services to innovation are:

- *Research Council of Norway*, which combines the roles of a traditional research council with that of an innovation agency, funding a broad spectrum of research from fundamental work in the universities to product and process development in companies.
- *Innovation Norway,* which handles business and regional development through a combination of loans, grants and advice.
- *SIVA,* the agency for infrastructure, investment and knowledge networks, as well as innovation centres, which owns and operates industrial and science parks, providing services from company incubation to renting property.

3.4.2.1. The Research Council

According to its mandate, the Research Council of Norway has three roles:

- Advise the government about research policy.
- Fund research.
- Create arenas for the actors of research, industry and government.

The RCN's strategy sets out six goals:

- To raise the quality of Norwegian research to the level of the other Nordic countries by 2010.
- Increase funding for research for innovation.
- Increase dialogue between research and society.
- Improve researcher careers to make research more attractive.
- Increase the internationalisation of Norwegian research.
- Improve RCN itself as a research council.

Box 3.3. The Research Council – a long-standing pillar of the Norwegian R&D funding system

In the immediate post-war era in 1946, Norway established first the NTNF, the Norwegian Research Council for Scientific and Industrial Research, linked to the Ministry of Trade and Industry (NHD). Within a couple of years, it began setting up various applied technology institutes – notably the Central Institute for Industrial Research (SI), which it owned and funded. In 1946, also, NLVF (the Norwegian Research Council for Agriculture) was established, linked to the Ministry of Agriculture (LD). A council for basic research was set up three years later, in 1949: NAVF, the Norwegian Research Council for Science and the Humanities. NAVF comprised four sub-councils: RMF, the Medical Research Council; RSF: the Social Science Research Council; RHF: the Research Council for the Humanities; and RNF: the Natural Sciences Research Council.

Throughout the post-war period the council system has played a key role in the Norwegian R&D policy debate. Several sectoral R&D initiatives were formulated as proposals to create new research councils, for example for consumer research and trade research. After a decade of debate, a research council for fisheries research was approved in 1972. A proposal by the Social Democrat government to create a council for research for societal planning in the mid-1970s resulted in the establishment in 1978 of a new sub-council within NAVF for research for societal planning (RFSP), which in 1987 became the independent NORAS council for the applied social sciences, while remaining under the responsibility of the Ministry of Education. Another semi-autonomous council for environmental research, NMF, was established within NAVF in 1987, and was generally considered to be the *de facto* research council of the Ministry for Environment.

…/…

Box 3.3. The Research Council – a long-standing pillar of the Norwegian R&D funding system *(continued)*

NTNF, and to some extent NAVF, began to assume tasks in addition to those for their "owner" ministries by the 1980s, foreshadowing RCN's later role as the servant of multiple ministries. A proposal put forward in 1970 to establish a single research council was turned down, owing to its "centralist" character. Several issues that arose during the 1980s, however, pointed to a need for reform in the research council structure.

It was emphasised throughout the 1980s that the research councils should be strategic agencies to a larger extent, that is, they mediate between the political level and the research-performing, institutional level. This issue was raised during the 1981 review of the NTNF system, in which the Norwegian system for industrial R&D was criticised for having become fragmented through 30 years of accumulation (not least of institutes) and through extensive earmarking of appropriations by the ministries. The strategic role of NTNF was also seen to conflict with its role as the legal owner of several research institutes, a criticism clearly inspired by the Rothschild reform in the United Kingdom, which institutionally separated customers and contractors in the public R&D system.

The latter half of the 1980s was a period of growth for Norwegian research. Most of the new public funds were channeled as appropriations to what was originally four growth areas, and which eventually became nine so-called main target areas *(hovedinnsatsområder)*. These were all cross-disciplinary and cross-sectoral research fields, and each main target area could involve several ministries and research councils: biotechnology; fishing and aquaculture (Havbruk); health, environment and the quality of life (HEMIL); information technology (IT); culture and research on the preservation and communication of traditions (KULT); management and organisation (LOS); oil and gas; materials technology; and environmental technology.

A complex committee structure was set up to cope with their cross-disciplinary nature, including national committees for strategy development, co-operation and co-ordination. A subsequent evaluation (Brofoss, 1993) argued that the existing research funding organisations contrived to fragment this national strategy and made it ineffective.

This was the starting point for the successive restructuring of the RCN leading to its present form.

The Council has a range of instruments and is organised into divisions for science, innovation, and strategic priorities. The Council had a budget of NOK 5.6 billion in 2007 (see Figure 3.10). Figure 3.11 shows that the institute sector is the largest beneficiary of RCN funding, closely followed by the university sector, reflecting the great importance of the institutes in the Norwegian knowledge infrastructure. The Council spends half of its budget on a very wide range of thematic and disciplinary research programmes; a quarter on infrastructural measures and 14% on bottom-up project funding (primarily in the Science Division) (Figure 3.12).

Figure 3.10. RCN budget, 2007

NOK millions

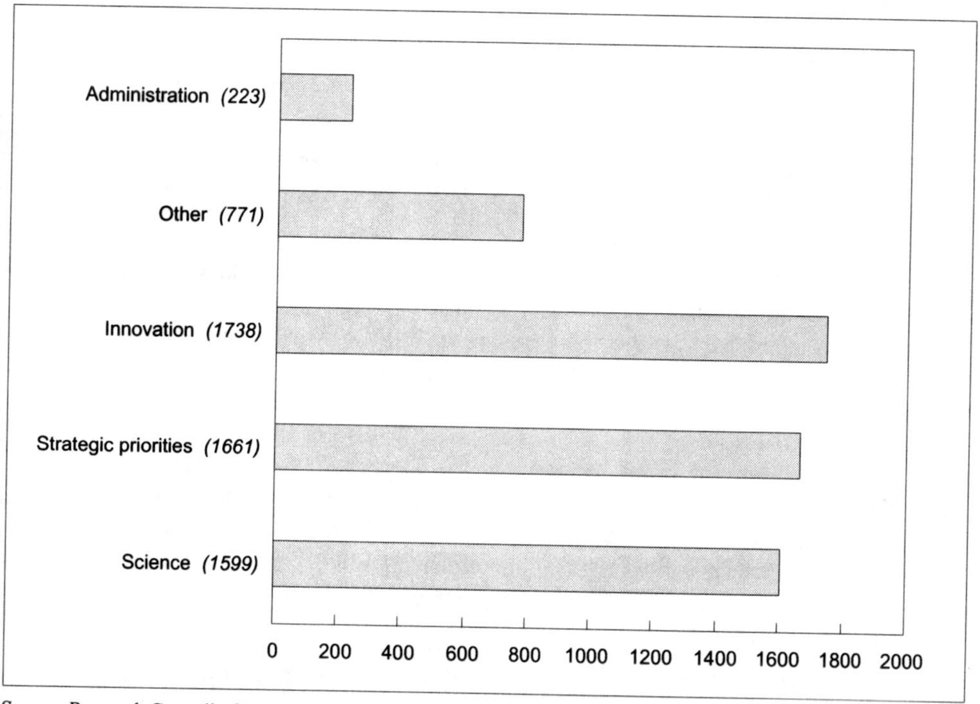

Source: Research Council of Norway.

Norway has gone through a process of proliferation and recentralisation of research and innovation funding agencies, and, while the current organisation probably works as well as can be expected, history and current practice show that in the presence of the strong sectoral principle and in the absence of an overall mechanism for policy debate and co-ordination, the best that can be achieved by centralising agencies is integration of policy implementation.

The Research Council – in its various forms – has been an important pillar of Norway's system of research funding since the immediate post-war period (see Box 3.3). Following a period of expansion of Norwegian research in the 1980s, the government had explained in its research White Paper (NOU 1988:28) that the research funding system had become complex and unworkable. Responsibilities were unclear and the government wanted to see a simplification of the entire structure based on a thorough analysis of existing structures. The Grøholt Committee was set up in 1990 to undertake this task, and reported a year later. Key points in its analysis were:

- The importance of governance mechanisms that make the research funding task clear and have legitimacy within the research-financing and research-performing communities.
- The unduly high influence of the sector principle, which hampered the development of a modern, internationalised research system and tended to suppress so-called free research.
- The prevalence of overlaps and inefficiencies in the funding system, not least the failure to separate strategic from operational issues and to delegate decisions to an appropriate level.

The Grøholt report recommended the creation of a single research council, with three disciplinary sub-councils, for life sciences, physical sciences and technology, and culture and social science. The merger in 1993 created an organisation with six divisions, which largely reproduced the old division of labour within a single organisation. The evaluation of the Council in 2001 criticised this as an expensive exercise in rearranging deck chairs that had largely failed to generate the increased co-ordination and synergy needed. It noted that the first two problems identified by Grøholt were still unresolved and that the third persisted in the form of detailed earmarking of research funding by the ministries – with the amount and detail of the earmarking generally being inversely related to the size of the ministries' research budgets.

In response, the Council was reorganised in 2002 into three operational divisions: science; innovation; and strategic priorities or programmes. The Innovation Division runs a large programme of user-directed research, which funds research done in-house by companies or (primarily) contracted by companies to external research performers, notably the Norwegian industrial research institutes. It has programmes supporting innovation in regional development and has newly introduced a programme of 14 Centres of Research-based Innovation, focused on research institutes. The Science Division handles fundamental and applied research. Like the rest of RCN, it makes strong use of programming. The share of RCN's spending on researcher-initiated free research rose from 9% to 12% between 2001 and 2006, the bulk of the increase being accounted for by the 13 centres of excellence launched in 2002 and the rest by programmes for big science and young researchers (Figure 3.10). Traditional response-mode project-based research increased from a little over NOK 300 million to a bit more than NOK 400 million over the period, but fell as a proportion of the budget from 9% to 8%.

Figure 3.11. RCN spending by beneficiary categories, 2002-06

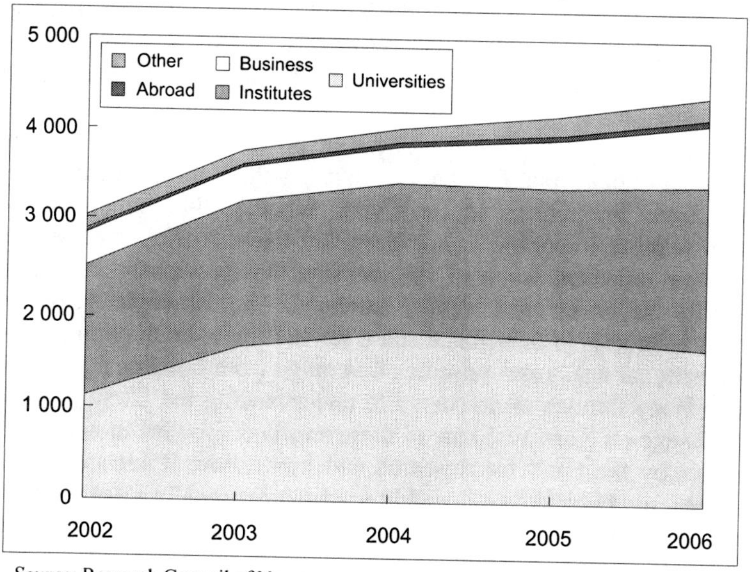

Source: Research Council of Norway.

Figure 3.12. Bottom-up funding of free basic research, 2001-06

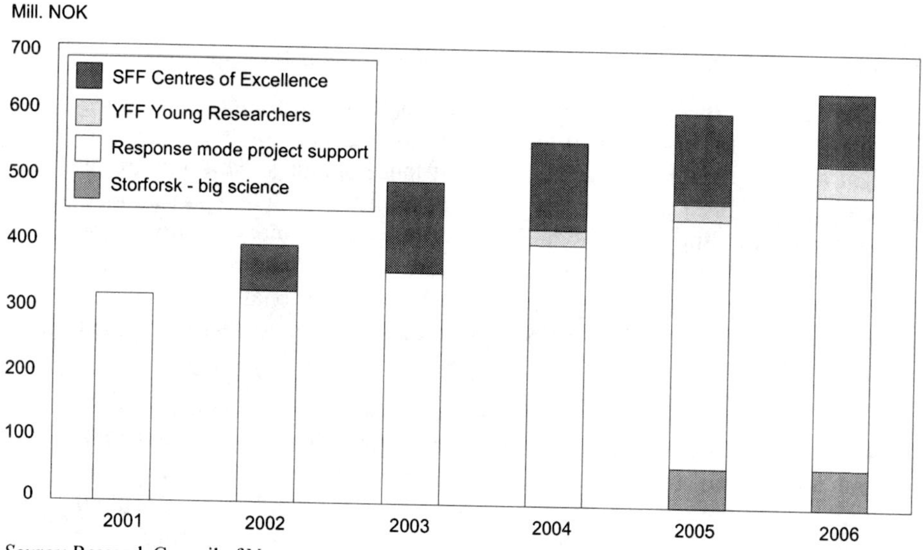

Source: Research Council of Norway.

Returns from the Fund for Research and Innovation enable the Research Council – which distributes a share of the yield – to launch new funding measures and enhance the financial strength of previously established schemes. For example, the fund is used to finance Research Council initiatives such as the Centres of Excellence and the Centres for Research-based Innovation, as well as a considerable portion of the Large-scale Programmes. In 2008, the Research Council will receive about NOK 940 million from the returns to the Fund. Whether the money or the organisation is more important is not clear, but together these two factors appear to have unlocked some of the lock-ins that previously hampered RCN's ability to create and deploy strategy. The Strategic Priorities Division has been able to launch and run a series of new large programmes focused on national needs and priorities that range from building a presence in nanotechnology through clean energy to understanding the likely impacts of climate change on Norway. Some of these are funded by the ministries in charge, others by the Fund for Research and Innovation. It seems to be a good practice – although not very widely used internationally – for research resources to be directed and redirected between more fundamental and more applied work according to need.

By receiving funds from a total of 16 ministries, RCN is an outstanding example of a "multi-principal agency" in research and innovation (Table 3.4). Each ministry earmarks the funding to varying degrees and, accordingly, one of RCN's administrative tasks is to produce an annual report in several parts to account for the use of each ministry's funding.

The Science Division is organised into departments for: social sciences, humanities, natural science and technology, biology and biomedicine, and clinical medicine and public health. Each of these operates via a mixture of research programmes (many at the behest of ministries) and response-mode or bottom-up funding of individual project proposals. Special funding for young researchers is available in this division, which is also responsible for the centres of expertise programme and for international scientific co-operation through network organisations such as ESF and facilities-based ones like CERN.

The Innovation Division operates the Council's user-directed R&D programmes (BIA), some branch-oriented programmes such as the MAROFF marine programme and the FIFOS programme for research in the public sector, the food programme, the new Centres for Research-based Innovation programme, the Skattefunn tax incentive and VRI, a programme for innovation in the regions that engages the capabilities of the regional universities, colleges and institutes. In addition, the division provides professional expertise to some of the programmes of the division for large programmes, for example in energy. There are also joint programmes with

Innovation Norway, for example for the wood industry (TRE), the Norwegian centres of expertise (modelled on the Finnish regional programme for providing technology support to smaller firms) and the FORNY commercialisation programme. The Innovation Division has responsibility for international innovation-related links, for example with the EU Framework Programmes and the TAFTIE network of European innovation agencies.

Table 3.4. RCN income by Ministry, 2005-06

	RCN income, 2006	Percentage of total
Ministry of Education and Research	1 279 603	24.4%
Ministry of Education and Research (excluding Fund for Research and Innovation	709 003	13.5%
Ministry of Trade and Industry	1 064 938	20.3%
Ministry of Petroleum and Energy	477 850	9.1%
Ministry of Fishing and Coastal Affairs	258 100	4.9%
Ministry of Culture and Food	362 763	6.9%
Ministry of the Environment	233 404	4.5%
Ministry of Children and Equality	16 140	0.3%
Ministry of Finance	8 500	0.2%
Ministry of Justice and Policey	8 000	0.2%
Ministry of Local Government and Regional Development	48 000	0.9%
Ministry of Culture and Church Affairs	12 661	0.2%
Ministry of Transport and Communications	133 200	2.5%
Ministry of Health and Care Services	174 764	3.3%
Ministry of Labour and Social Inclusion	93 200	1.8%
Ministry of Government Administration and Reform	10 218	0.2%
Ministry of Foreign Affairs	131 452	2.5%
Administrative budget (from the Ministry of Education and Research)	217 479	4.2%
Total all ministries	5 239 275	100%

The Strategic Priorities Division has five thematic departments and currently runs major programmes related to national priorities: VERDIKT: ICT; FUGE: functional genomics (see Box 3.4); NANOMAT: nanotechnology and materials (see Box 3.4); PETROMAKS: research to maximise the exploitation of oil deposits; RENERGI: future clean energy technologies; NORKLIMA: climate change and its implications for Norway.

It also co-ordinates some larger programmes that work across divisions, bringing together basic research and innovation.

Box 3.4. The FUGE and NANOMAT programmes

FUGE – Functional genomics in Norway

FUGE was launched in 2001 with the main objective to bring research in functional genomics up to an international level within a five-year period. Prioritised areas were marine, research, medical research and basic biological research – including plants, animals and micro-organisms.

The major task in the first phase of FUGE (2002-06) was to establish technology platforms and/or develop world-class research groups. The platforms were: biobanks for health, bioinformatics, integrative genetics, arabidopsis research centre, microarray technology, structural biology, molecular imaging, transgenic mice, microbial technology, proteomics, and a regional research biobank in central Norway. An evaluation of the first phase has recently taken place.

The next phase (2007-11) builds on the somewhat modified platform concept as technologies have matured and become more widespread and as other needs have appeared. More emphasis is put on industrial applications. Still the thematic concentration will be on medicine and the marine area although agricultural and industrial aspects will also be dealt with. The entire FUGE programme is foreseen to have overall funding of approximately NOK 1.5 billion for the period 2002-11.

NANOMAT

According to the annual R&D Survey, research in nanotechnology is a small but non-negligible activity in the private sector, amounting to NOK 135 million (1% of total R&D costs). The bigger spenders are found in the materials and petroleum sectors.

The Research Council has recently presented its proposed N&N strategy to the Minister for Education and Research. This strategy prioritises four thematic areas: energy and environment, ICT and microsystems, health and biotechnology, and sea and food. In addition it focuses on the following areas of competence: materials; surface/interface science and catalysis; fundamental physical and chemical processes at the nanoscale; bio-nanoscience and biotechnology; devices, systems and complex processes based on N&N; ethical, legal and societal aspects (ELSA) including health, safety and environment. The strategy also recognises essential tool platforms for N&N: synthesis, manipulation and fabrication; characterisation; and theory and modelling. ELSA should be integrated into relevant projects and centres, and should be co-ordinated with similar research for other technologies.

Based in the Research Council, the NANOMAT (*Nanoteknologi og nye materialer*) programme will run from 2002 to 2011 and spans research in nanotechnology and in materials. For 2007, the budget for NANOMAT is NOK 67.5 million.

In addition to the programmes of its three operational divisions, RCN has responsibility for channelling core funding to the 61 research institutes covered by the state regulations for financing and for evaluating institutes. To date, the evaluations have been well conducted but have at least no direct effect on the institutes' incentives, which continue to be set by the ministries. A proposed a new, indicator-based way for determining the institutes' core funding would alter this, if implemented (in 2009) and previous evaluations seem to have been instrumental in triggering this innovation.

In practice, RCN's ability to design and implement strategy – and therefore to play the potential co-ordinating role implied by its position as a multi-principal agency –may be constrained by the ministries that specify in detail how the Council is to use their funds. The Council has been able to reduce fragmentation to some extent by increasing mean project size and duration in recent years (especially in the Innovation Division) and by setting up the Centres of Excellence and Centres for Research-based Innovation. The appropriations to the Research Council from the Ministry of Education and Research, the Ministry of Trade and Industry, and the Ministry of Petroleum and Energy are generally not earmarked at a detailed level. The appropriations from these ministries account for 67% of the RCNs total budget (see Table 3.4). The earmarking that is done by these ministries is mainly in line with the RCN's own priorities in their budget proposals or are a result of clear political priorities (*e.g.* the establishment of the CoE scheme, the YFF scheme and the FUGE programme, none of which were originally initiated by the RCN).

The Ministry of Education and Research recently asked RCN to undertake a self-evaluation as a way to review its progress since the 2003 reorganisation, and concluded that while it has made progress in becoming more open and in providing better service to its customers, it was still not adequately able to co-ordinate across divisions, for example by establishing matrix structures. At the same time, it pointed out that the reorganisation – which reduced the number of operating divisions from six to three – Science, Innovation and Strategic Priorities – had created a division for basic research and improved the quality of the dialogue between the Council and its various stakeholder groups.

3.4.2.2. Innovation Norway

Innovation Norway is "owned" by the Ministry of Trade and Industry and functions as a one-stop shop for business and innovation support. It provides an impressive and quite complete set of services, largely to SMEs, and offers a strong national infrastructure able to design and deploy instruments that support not only national but also regional development policy. Its mixed national and regional governance structures are complex.

In principle, with the agency's good relationship with RCN, they allow for the creation of mutually supportive initiatives that can promote innovation at both regional and national levels. Given the size and complexity of the organisation, it would require a substantial evaluation to assess its overall effectiveness and role.

Like RCN, Innovation Norway is the result of a series of mergers dating back to the 1990s. In 1993, SND was created through a merger of the Industry Fund *(Industrifondet)*, which had served as a business development agency, the Industry Bank *(Industribanken)*, which had a nationwide mandate as an industrial development bank, the SME Fund (*Småbedriftsfondet*), which had a similar mission for small companies, and the regional development fund *(Distriktenes Utviklingsfond)*, which functioned as a regional development bank. In 1996, the specialised Fisheries Bank *(Fiskarbanken)* was also integrated into SND. SND became a wide-ranging combination of a development bank and business development agency, offering a wide range of loans and advisory services as well as some grant-aid schemes in areas such as management training, innovative procurement and company networking. The main banking operations were profitable and the income was used to supplement SND's budgets from the Trade and Industry and Regional Development ministries. Its SND Invest Division acted as an accessible source of venture capital able to take comparatively high risks because its owner set a low target in terms of rate of return compared with the private sector. SND had special responsibility for supporting regions experiencing economic difficulties and had a small group of consultants who could be allocated by the Ministry for Regional Development to communities needing special measures. SND had responsibility for encouraging FDI, but this function was largely dormant by the late 1990s.

SND operated through a national network of 18 offices – in effect, one per county – to provide a one-stop shop for company development in partnership with the counties, whose representatives (especially the business and innovation development functions) shared the governance of the regional offices with the central state. In order to simplify the support infrastructure and to strengthen the links between regional development, innovation and internationalisation (St. ptp. nr. 51 2003-2003), in 2004 SND was merged with the Export Council, the inventors' advisory organisation (SVO) and the Norwegian Tourist Board to form Innovation Norway. The merger with the Export Council meant the organisation acquired a further 38 offices in 31 countries in addition to being close to users through SND's network of regional offices.

The law establishing Innovation Norway states that it shall develop and implement instruments that contribute to increased innovation by companies in all parts of the country and act as the state's primary agency for company-oriented programmes. Customers are to be central to its activities. The main target groups are entrepreneurs, young companies and SMEs with growth potential. Innovation Norway should promote female entrepreneurship and actively develop innovation systems. The goal of making Norwegian business internationally competitive should influence all Innovation Norway's activities.

Innovation Norway has formulated its mission as: to promote economically and socially profitable business development in the entire country and to enable regions to realise their business potential by contributing to innovation, internationalisation and marketing. Its motto is "We give local ideas global opportunities". Innovation Norway has four divisions as well as tourism and general staff functions. The divisions are: Norway; International; Services and Programmes; Strategy and Communications.

While it is "owned" by the Ministry of Trade and Industry, Innovation Norway acts as an agency for three ministries. It allocated total funding of NOK 4.6 billion in 2006, of which 44% were Ministry of Trade and Industry (NHD) funds' 30% were Ministry of Regional Development (KRD) funds; 26% were Ministry of Food and Agriculture (LMD) funds; and 1% Ministry of Fisheries and Coastal Affairs (FKD) funds.

The KRD essentially uses Innovation Norway to channel funds to the counties, both grant and loan support for their regional development activities and transfers in the form of transport and employment tax subsidies for the more remote regions (together these accounted for 9% of Innovation Norway's allocations in 2006). NHD largely funds national innovation and internationalisation-related activities, while the LMD funding focuses on rural development. FKD focuses on coastal areas.

Some 35 percentage points of the Ministry Trade and Industry funds were allocated in the form of loans rather than grants. In all, 40% of the funds allocated were loans. Some 33 percentage points of the total funds allocated were so-called "low risk loans" – because of Innovation Norway's, and SND's before it, track record of operating a profitable and successful development banking business with a low level of default – and a further 7 percentage points were riskier loans.

Innovation Norway reports that 11% of its 2006 allocations went to start-ups and a further 87% went to support innovation in established firms, of which about 33 percentage points were for "young businesses", "growth SMEs", spin-offs etc. Only 3% of the total budget was reported to be for internationalisation activities, but a significant part of the international division's activities is paid for by Innovation Norway's customers. Innovation Norway has a comprehensive system of performance indicators, which it reports annually, along with the reports of its customer satisfaction survey. However, the organisation's reporting clearly shows that it has three principals, each of which is quite specific about how it wants its money spent, a reflection of the strength of the "sector principle" in Norway.

Innovation Norway offers a very wide range of services. It currently focuses on 14 major areas of "effort" (*satsing*), using multiple services and programmes for each. These comprise eight industrial branches plus network-based innovation, SMEs with international growth potential, start-ups, female entrepreneurship and young people's entrepreneurship. It operates 19 programmes and offers a total of 38 services or instruments.

The services offered span a wide range of loans, grants and guarantees; advice; training; export market information, partner search, publicity; branch support programmes; networking; mentoring and support through non-executive board members; special support programmes for farming and fishing, including help with income diversification; and so on. The list is as impressive for its length as for its breadth.

Given that Innovation Norway funds commercialisation, start-up and innovation, it is not surprising that there is some overlap with RCN and that the two organisations share some activities. Through agreements with RCN, Innovation Norway shares in the management of the FORNY commercialisation programme as well as the Norwegian centres of expertise and the ARENA programme which aim to promote regional innovation in clusters (see OECD, 2007d, Chapter 16). RCN representatives are available in eight of the Innovation Norway district offices and the others are equipped to refer companies to RCN services, where appropriate.

3.4.2.3. SIVA (Selskapet for Industrivekst)

SIVA (the Company for Industrial Growth) was set up in 1968 to develop industrial estates in rural areas of Norway, but expanded its role from about 1990 into the nationwide provision of other types of parks to provide space and services to developing companies and venture capital to start-ups in its own facilities. SIVA is currently involved in: 22 incubators; 18 knowledge parks, which are effectively small science parks attached to regional colleges; 50 business gardens or small industrial estates; eight research parks, which are more or less traditional science parks; and has an interest in nine regional venture and seed funds. It operates a number of company networks. It provides advice and mentoring to its tenants and can support them by placing experienced people on their boards of directors.

SIVA was evaluated in 2000 by STEP, which concluded that it was a capable and profitable organisation that overlapped little with the others and had a useful and positive effect on employment. However, it was less successful in the regions (where its primary responsibilities lie) than in large cities and had had only a limited effect on establishing value-creating milieus. STEP found that it was spread too thin and suggested that it should concentrate its efforts to a greater extent. It would be fair to say that SIVA has since done the opposite.

Table 3.5. Public purchases in Norway, 2004-05

NOK billions

	2004	2005	Change (%)
Total	257.0	276.2	7.5
General government	203.9	210.4	3.2
Central government	120.8	125.3	3.7
Central government excluding defence	103.7	108.7	4.8
Defence affairs and services	17.1	16.6	-2.9
Local government	83.1	85.1	2.5
Municipalities	72.9	74.0	1.5
Counties	10.2	11.1	9.4
General government enterprises	53.1	65.8	24.0
Central government excl. oil	9.3	7.7	-18.0
Oil sector	37.9	53.9	42.4
Local government	5.9	4.2	-28.1

Source: Background report.

3.4.2.4. Demand-side policy: the role of public procurement and innovation in public services

There is reviving international interest in demand-side innovation policies based on procurement. Hambro (2007) proposes that Norway should increase the scale of its long-running and successful OFU programme for supporting innovative procurement. He points out that state procurement is usually far from innovative, because it favours low-risk, off-the-shelf solutions and says that this culture needs to be changed, for example by instituting performance indicators that encourage innovative procurement, by moving towards more multi-annual budgeting in order to recognise not only the short-term capital costs of innovation but also the longer-term cost savings, by providing subsidies and guarantees for innovative procurement and by establishing a powerful organisation to promote changes in state procurement practice.

Table 3.5 certainly suggests that there might be scope for such action. Edler and Georghiou (2007) point out that, in the context of WTO rules as well as in the interest of obtaining efficient solutions, a new generation of demand-side policies needs to focus on defining markets and needs and to avoid betting on predetermined national champions or their equivalents.

Even if Norway has a rather scattered and decentralised system of public procurement in which each public agency is responsible for its procurement activities and collaboration, there are examples of collaboration and professionalism among the central bodies. Still there is lot to be done if Norway wants to have the appropriate skills in all departments responsible for public procurement. Current public procurement skills lie more in efficiency than in stimulating research and innovation. For example, procurers are measured according to several criteria, but research and innovation are not among them.

Norway has in fact been active in this area for some time. The OFU/IFU programme was transferred to SND from the Ministry of Trade and Industry in 1994-95 and is today operated by Innovation Norway. Originally a programme to use state demand to stimulate the development and production of innovative products (OFU), it was extended to using industrial buyers as motors for innovation in 1994 (IFU). A recent evaluation found that the schemes were effective and yielded significant socio-economic benefits (Econ, 2007). Unlike many of SND's other measures, which deal with innovations that are new to the firm, OFU/IFU focuses on innovations which are new to the world. However, as Hambro suggests, there are opportunities to extend the use of this potentially powerful approach.

In recent years it has been recognised that – given the importance of the public sector in Norway – increasing the rate of innovation in public services should have significant social and economic benefits.

3.4.2.5. A summary diagnostic

As Norwegian policy makers have over the years been very active in innovating but also in imitating research innovation support instruments, the portfolio is broad. There appears to be scope for more demand-side measures, as is recognised in current policy discussions. Similarly, given the scale of the public sector, measures to promote innovation in that part of the economy would be useful. The services sector is economically important and authorities should make sure that it receives the attention it deserves in research and innovation policy.

Between RCN, Innovation Norway, SIVA and the state's seed and venture capital activities, Norway has a fairly complete set of instruments to support research and innovation. Many of these are home-grown, especially in the innovation support area where user-directed R&D (BIA) at RCN and the FRAM strategy training programme for small companies at Innovation Norway stand out as examples of especially good practice. The Norwegian system is also willing to learn from and adapt international experience. The centres of excellence are based on international models. The competence centres were designed on the basis of studies of foreign competence centres programmes, which were adapted to the Norwegian knowledge infrastructure, with its large institute sector. The Norwegian centres of expertise have international precedents (OECD, 2007d, Chapter 16).

The proportion of RCN support for R&D that goes to industry is quite low. Programmes such as NANOMAT, which aim to build national capacity in research and industry, appear to underexploit opportunities for more active linkages between industry and the knowledge infrastructure by using user-directed R&D funding to involve industry rather than providing support to activity inside the firms and building links from this more active company participation. In this respect, closer examination of international good practices (*e.g.* Tekes' technology programme model which links academic and industrial research sub-programmes in areas that are prioritised as a result of extensive consultation with both academia and industry) may be valuable.

Finally, the amount of resource devoted to demand-side policy, compared with the bulk of RCN's spending (which targets manufacturing and to a lesser extent the primary sector – together a rather small part of the Norwegian economy) – is very modest. At present, RCN appears to run no research or innovation programmes specifically oriented to the service industry. Innovation Norway, being the primary agency for programmes aimed at business firms, in particular young companies and SMEs with a high growth potential, is also active in this area. Innovation Norway has programmes and efforts directed towards service industries (travel, culture and catering).

Annex A

NORWAY'S R&D PERFORMANCE IN INTERNATIONAL PERSPECTIVE: BENCHMARKING INDICATORS

At least two factors complicate international comparison of Norway's R&D spending and performance. Such comparisons are usually made by normalising against GDP, but because Norway has an unusually large GDP owing to its oil revenues, GDP-based comparisons tend to show Norway in an unfavourable light compared to countries that lack such large natural resources. Usually, the tendency is to compare a country to its neighbours and it is especially marked in the Nordic area because of historical, cultural and economic ties. However, Sweden, Finland and Iceland not only have different industrial structures from Norway but also invest very heavily in R&D, with Sweden and Finland typically at the top of various international benchmarks of national R&D performance.

Figure A1.1 shows an apparent relationship between R&D intensity and GDP per capita – though there is enough dispersion around the trend to show that there are other factors at play. Norway is clearly wealthier than would be expected from its R&D investment, while Sweden and to a lesser degree Finland are considerably poorer.

Table A1.1 shows that Norway is the Nordic country that spends the smallest proportion of GDP on R&D – below both the EU15 and OECD averages. This is partly a reflection of its large GDP. In money terms, per capita Norwegian R&D expenditure exceeds the international average but still trails the other Nordic countries.

Figure A1.1. R&D intensity (GERD/GDP) and wealth (GDP per capita), 2004

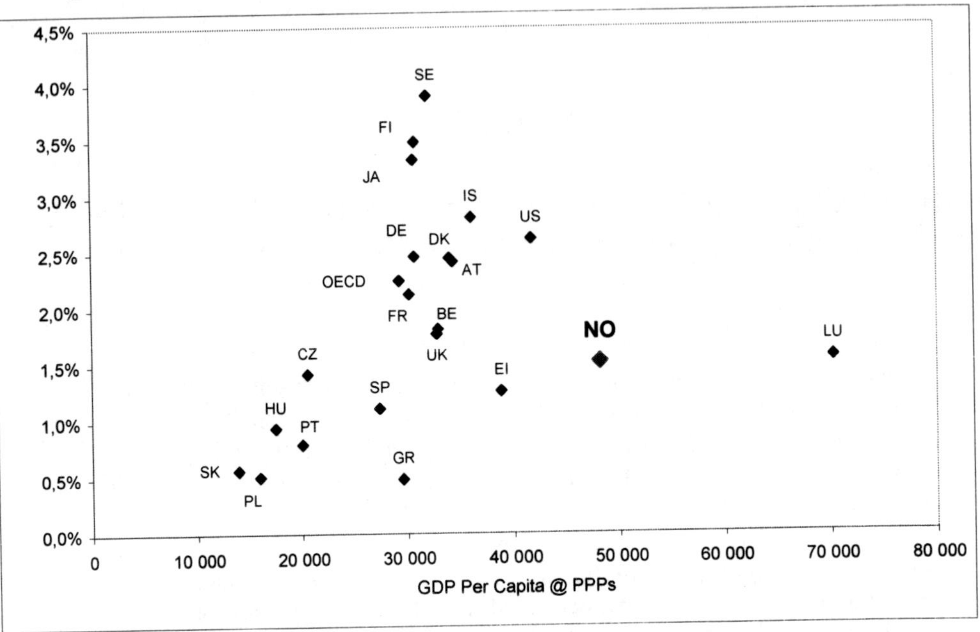

Source: OECD Main Science and Technology Indicators, 2007.

Table A1.1. Key indicators for R&D in Nordic countries, EU27 and OECD, 2005

	Norway	Sweden	Denmark	Finland	Iceland	EU (27)	OECD
GERD/GDP (%)	1.52	3.89	2.45	3.48	2.78	1.74	2.25
GERD per capita (USD PPP)	735	1248	834	1077	1017	472	661
% of GERD financed by government	44	24	28	25	41	35	30
GBAORD per capita (USD PPP)	343	286	243	318	348		

Source: OECD Main Science and Technology Indicators, 2007.

Figure A1.2. Nordic countries' gross investment in R&D, 1981-2005

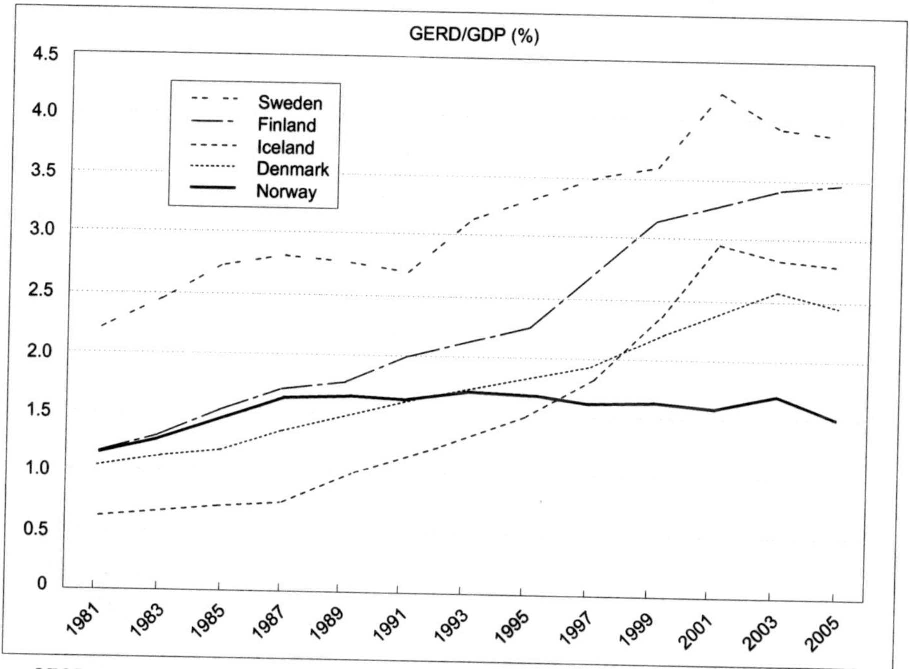

Source: OECD Main Science and Technology Indicators, 2007.

Figure A1.2 shows Norway's GERD stagnating as a proportion of GDP, while the other Nordic countries surge ahead. However, a comparison with the major industrial countries and the EU (Figure A1.3) suggests a different story, with Norwegian investment bumping along just below, but more or less in parallel with, that of other countries, with Japan pulling away from the pack.

Figure A1.3. Norwegian GERD/GDP compared with the largest OECD countries

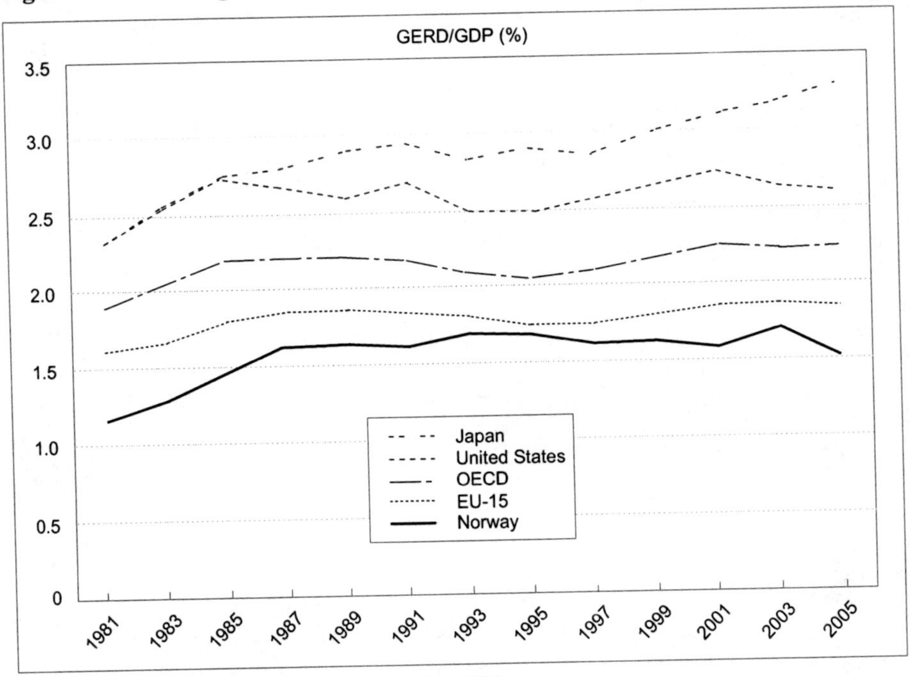

Source: OECD Main Science and Technology Indicators, 2007.

Figure A1.4. State vs. business-funded R&D as a proportion of GDP, 2005

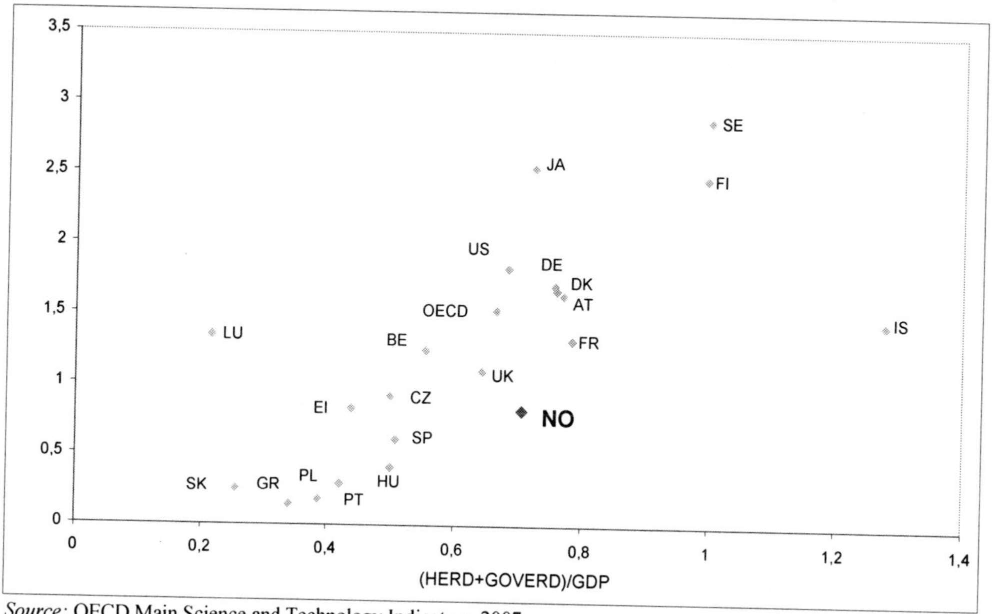

Source: OECD Main Science and Technology Indicators, 2007.

Figure A1.4 shows an apparent relationship between the proportion of GERD devoted to BERD and that spent on the knowledge infrastructure, suggesting that the public and private components of GERD need to be kept in balance. Again, there is a fair amount of dispersion around the trend suggesting that other factors are at work, but the figure supports the idea that the two kinds of R&D are to some degree interdependent and that the knowledge infrastructure is exactly that: an infrastructure that makes BERD possible, not least because of its role in producing the necessary human capital. This does not imply that it is possible to drive up BERD by investing more in the knowledge infrastructure – but it does suggest that insufficient investment in the knowledge infrastructure can impede growth in BERD. At the level of total R&D, however, the high ratio of state to business investment in R&D suggests that there is little risk of this in Norway. If there is a problem in the raw quantity of Norwegian GERD, it is more likely to lie in business than in state expenditure. Norway's position on Figure A1.4 implies an unusually high ratio of state to business expenditure on R&D.

Figure A1.5. Nordic countries' state spending on R&D, 1981-2005

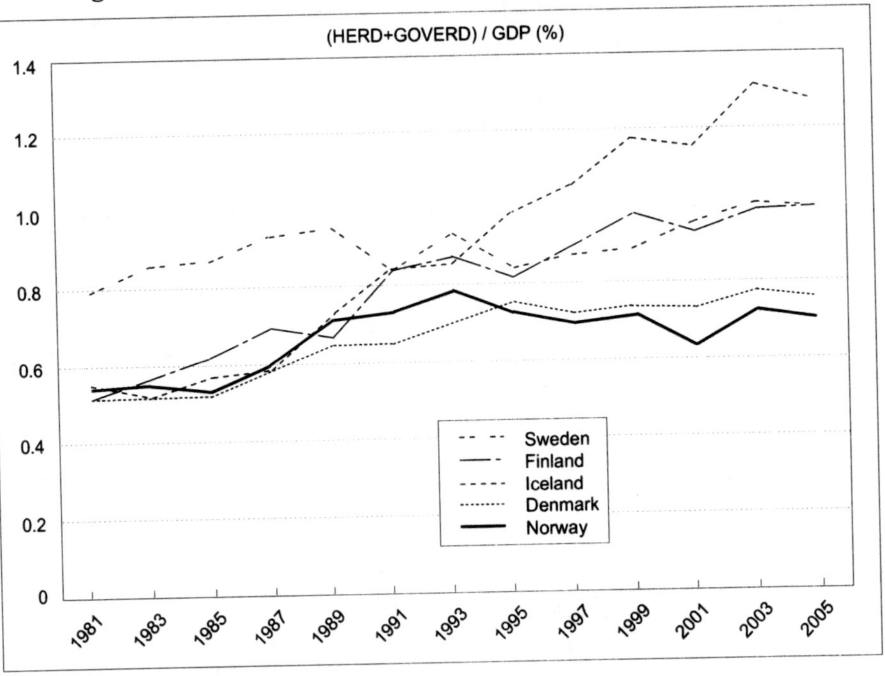

Source: OECD Main Science and Technology Indicators, 2007.

The Barcelona goal means that state spending on R&D should reach 1% of GDP and BERD 2% of GDP. In the short to medium term, it is not obvious that Norwegian state R&D investment needs to increase unless BERD is in some sense under-counted in Norway. Figures A1.5 and A1.6 imply that the effort required for either Norway or the EU15 to increase state spending to the required level would be unfeasibly large, even though the figures also show a Nordic pattern of comparatively high state spending on research.

Figure A1.6. Norwegian state spending on R&D compared with the largest OECD countries, 1981-2005

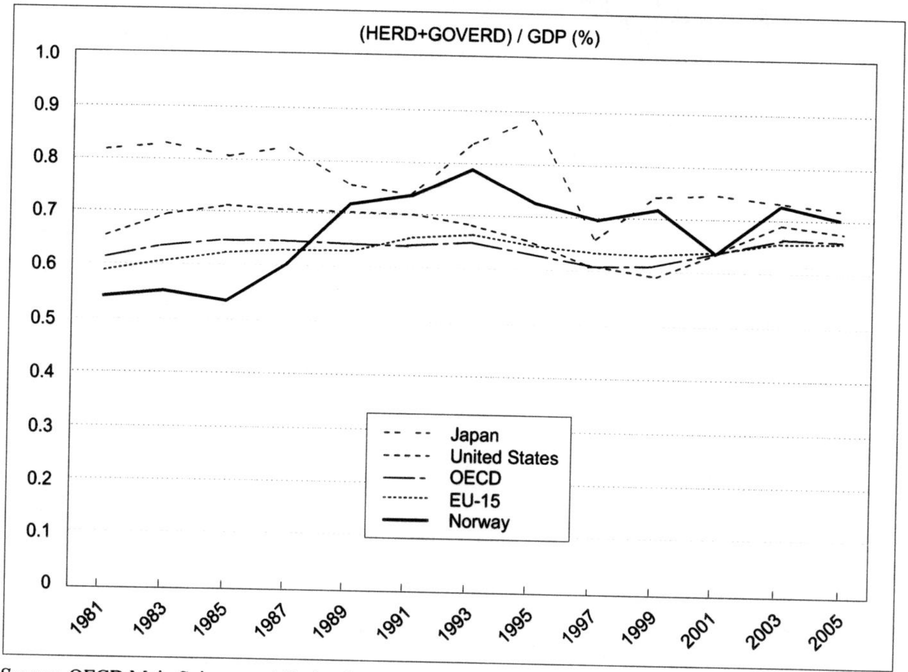

Source: OECD Main Science and Technology Indicators, 2007.

Figure A1.7. Nordic countries' GERD per capita, 1981-2005

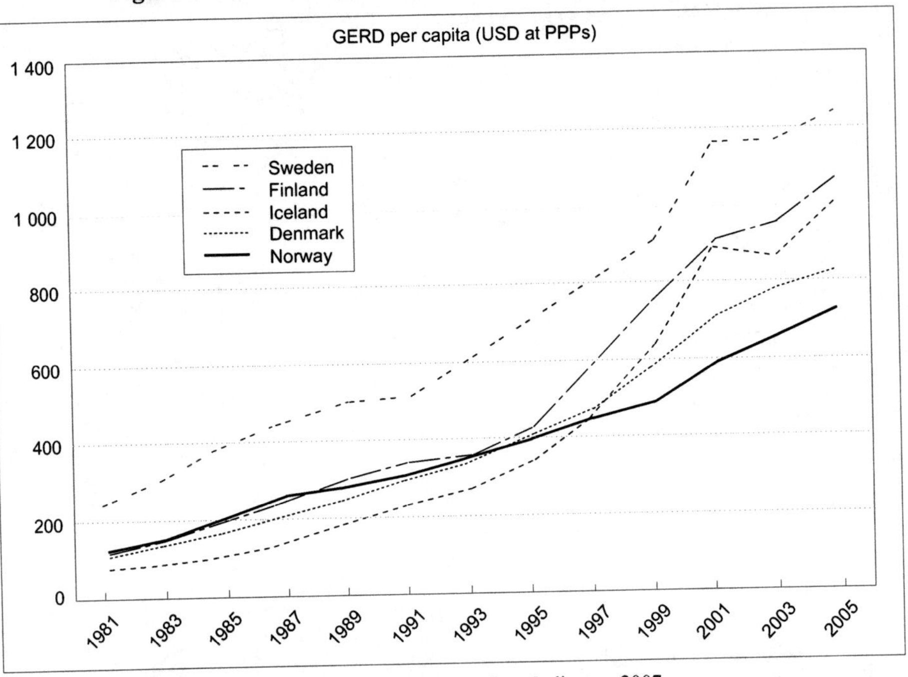

Source: Calculated from OECD Main Science and Technology Indicators 2007.

However, Norway's comparative position in R&D investment and performance looks rather different when it is normalised by population rather than GDP. Compared to other Nordic countries (Figure A1.7) the Norwegian rate of R&D spending is lower but increasing, a marked contrast to the stagnation in Norwegian GERD when it is normalised by GDP. Similarly, if the GDP effect is removed, Norwegian R&D investment per capita tracks that of the EU15 and the OECD and looks quite respectable.

Figure A1.8. Norwegian GERD per capita compared with the largest OECD countries

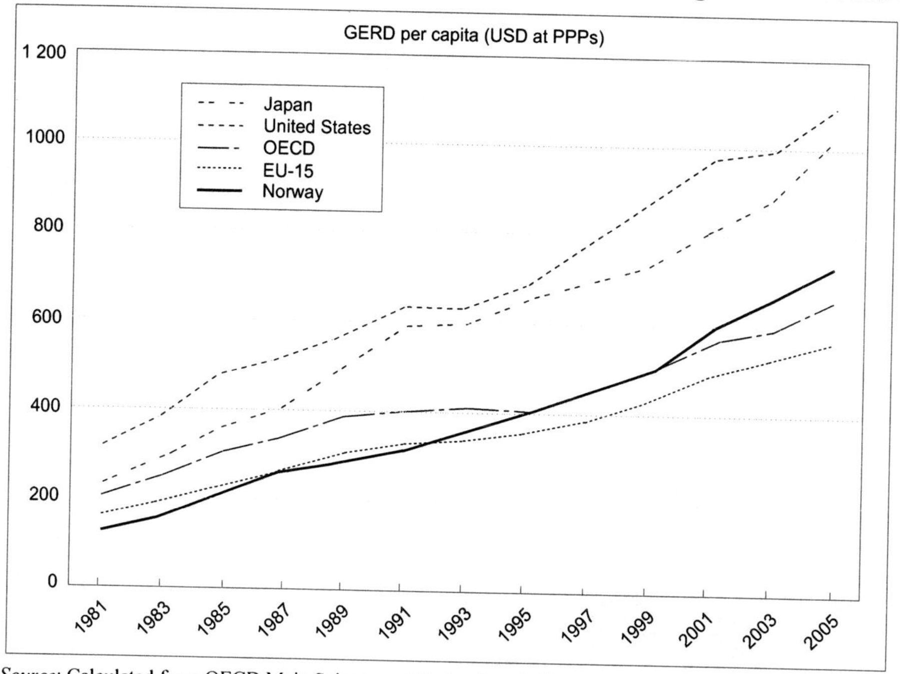

Source: Calculated from OECD Main Science and Technology Indicators 2007.

References

Aghion, P. and P. Howitt (2005), "Appropriate Growth Policy: A Unifying Framework", Joseph Schumpeter Lecture, August.

Ahn, S. (2001), "Firm Dynamics and Productivity Growth", OECD Economics Department Working Paper No. 297, OECD, Paris.

Ahn, S. (2002), "Competition, Innovation and Productivity Growth: A Review of the Theory and Evidence", OECD Economics Department Working Paper No. 317, OECD, Paris.

Aiginger, K. (2006), "Competitiveness: From a Dangerous Obsession to a Welfare Creating Ability with Positive Externality", *Journal of Industry, Competition and Trade*, 6, pp.161-177.

Arnold, E., S. Kuhlmann and B. van der Meulen (2001), *A Singular Council: Evaluation of the Research Council of Norway*, Ministry of Education, Research and Church Affairs, Oslo.

Arnold, E. and P. Boekholt (2003), "Research and Innovation Governance in Eight Countries: A Meta-Analysis of Work Funded by EZ (Netherlands) and RCN (Norway)", Technopolis, Brighton (available at www.technopolis-group.com).

Arnold, E. (2004), "Evaluating Research and Innovation Policy: A Systems World Needs Systems Evaluations", *Research Evaluation*, Vol. 13 No. 1, April.

Arnold, W., N. Brown, A. Eriksson, T. Jansson, A. Muscio, J. Nählinder and R. Zaman (2007), "The Role of Industrial Research Institutes in the National Innovation System", VA 2007:12, VINNOVA, Stockholm.

Arnold, E., J. Clark and A. Muscio (2005), "What the evaluation record tells us about Framework Programme performance", *Science and Public Policy*, Vol. 32, No. 5, pp. 385-397.

Arnold, E., A. Eriksson, S. Faugert and T. Jansson (2006), *Building Nordic Strength Through More Open R&D Funding*, TemaNord 2006:576,: Nordic Council of Ministers, Copenhagen.

Arrow, K. (1959), 'Economic Welfare and the Allocation of Resources for Invention,' reprinted in Richard Nelson (ed.), (1962), *The Rate and Direction of Inventive Activity*, Princeton University Press, New Jersey.

Bellone B. and A. Bibbee (2006), "The Ageing Challenge in Norway: Ensuring a sustainable pension and welfare system", Economics Department Working Paper No 480, OECD, Paris.

Bernt Commission (2001), NOU 2001:11, *From Insight to Industry (Fra innsikt til industri)*.

Bosma, N. and R. Harding (2006), "Global Entrepreneusrship: GEM 2006 Summary Results", Babson College, Babson, ME.

Brofoss, K.-E. (1993), Innsatsområdene som forskningspolitisk virkemiddel, report 4/1993, NIFU, Oslo.

Cadiou, Y., E. Arnold, M.-L. Eriksson and B. Good (forthcoming 2008), *Nordic Excellence: A Bibliometric Exploration of Common Nordic Research Funding Opportunities*, Nordforsk, Oslo.

Edler, J. and L. Georghiou (2007), "Public procurement and innovation – Resurrecting the demand side", *Research Policy*, Vol. 36, pp. 949-963.

Emblem, T. (1995), "Strategi for instituttsektoren. Mål, struktur, organisering, Rapport nr 3 fra prosjekt om instituttpolitikk i Norges forskningsråd", Forskningsrådet, Oslo.

Ernst & Young Management Consulting (1998), "Evaluering av myndighetenes deltaking i regionale såkorn- og venturefond", Oslo.

European Commission (2005), *Public Procurement for Research and Innovation, Expert Group Report on Developing Procurement Practices Favourable to R&D and Innovation,*: DG Research, Brussels.

Freeman, C. (1987), *Technology and Economic Performance – Lessons from Japan*, Pinter Publishers, London.

Hambro, C. (2007), *Offentlige anskaffelser som fremmer innovasjon: Analyse og forslag til tiltak, En utredning for NHO*, Gram, Hmbro & Garman, Oslo.

Høj, J., M. Jimenez, M. Maher, G. Nicoletti and M. Wise, "Product Market Competition in the OECD Countries: Taking Stock and Moving Forward", OECD Economics Department Working Paper No. 575, OECD, Paris.

Kallerud, E., J. Hauknes and P. Koch (2006), *European Trend Chart on Innovation: Annual Innovation Policy Trends and Appraisal Report*, Norway, 2006 at http://trendchart.cordis.europa.eu/.

Kaloudis, A. and C. Rørstad (2006), "Analysis of Public R&D Funding in Norway", NIFU STEP, Oslo, downloadable from www.nifustep.no.

Kline, S.J and N. Rosenberg (1986), "An Overview of Innovation", in R. Landau and N. Rosenberg (eds.), *The Positive Sum Strategy: Harnessing Technology for Economic Growth*, National Academy Press, Washington, DC.

Koch, P. and J. Hauknes (2007), "Understanding the Norwegian Puzzle", (mimeo) RCN, Oslo.

Lucking, B. (2004), "International Comparisons of the Third Community Innovation Survey", Department of Trade and Industry, London.

Lundvall, B.-Å. (1988), "Innovation as an Interactive Process: from User-producer Interaction to the National system of Innovation"; in G. Dosi, C. Freeman, R. Nelson, G. Silverberg and L. Soete (eds.) (1988), *Technical Change and Economic Theory*, Pinter Publishers, London and New York.

Lundvall, B.-Å. (ed.) (1992), *National Systems of Innovation: Towards a Theory of Innovation and Interactive Learning*, Pinter Publishers, London.

Martin, B. and J. Irvine (1989), *Research Foresight: Priority-Setting in Science*, London: Pinter Publishers, London and New York.

Martin, B. (1995), "Foresight in Science and Technology", *Technology Analysis & Strategic Management*, Vol. 7, No. 2, pp. 139-168.

Martin B. and R. Johnston (1999), "Technology Foresight for Wiring Up the National Innovation System - A Review of Recent Government Exercises", *Technological Forecasting and Social Change*, 60 (1), January, pp. 37-54.

Metcalfe, S. (1995), "The Economic Foundations of Technology Policy: Equilibrium and Evolutionary Perspectives", in P. Stoneman, (ed.), *Handbook of the Economics of Innovation and Technical Change*, Blackwell, London.

Mowery, D.C. and N. Rosenberg (1978), "The Influence of Market Demand upon Innovation: A Critical Review of Some Recent Empirical Studies", *Research Policy*, April 1978.

Narula, R. (2002), "Innovation Systems and 'Inertia'", in "R&D Location: Norwegian Firms and the Role of Systemic Lock-in", *Research Policy*, Vol. 31, Issue 5, pp. 795-816.

National Science and Technology Policy Council of Finland (2003), *Knowledge, Innovation and Internationalisation*, Ministry of Education, Helsinki.

Nelson, R.R. (1959), "The Simple Economics of Basic Scientific Research",' *Journal of Political Economy*, Vol 67, pp 297-306.

Nelson, R.R. (1988), "Institutions Supporting Technical Change in the United States", in G. Dosi, C. Freeman, R. Nelson, G. Silverberg and L. Soete (eds.) (1988), *Technical Change and Economic Theory*, Pinter Publishers, London and New York.

Nelson, R.R. (ed.) (1993), *National Innovation Systems: A Comparative Analysis*, Oxford University Press, New York and Oxford.

NESTA (2006), "The Innovation Gap: Why Policy Needs to Reflect the Reality of Innovation in the UK", NESTA Research Report, October, National Endowment for Science, Technology and the Arts, London.

NHD (2003), *From Idea to Value: The Government's Plan for a Comprehensive Innovation Policy*, NHD Oslo.

NIFU, STEP and Technopolis (2004), *Evaluation of Norway's Participation in the EU's Fifth Framework Programme*, NIFU STEP, Oslo.

Norwegian Ministry of Education and Research (2006), *Strategy 2006-2009: A Joint Promotion of Mathematics, Science and Technology.*

NOU (1981), *Research, Technological Development and Industrial Innovation* (Forskning, teknisk utvikling og industriell innovasjon), 1981:30.

NOU 1988:28, Med viten og vilje.

OECD (1999), *Managing National Innovation Systems*, OECD, Paris.

OECD (2002), *Dynamising National Innovation Systems*, OECD, Paris.

OECD (2003), *Learning for Tomorrow's World. Trends in International Mathematics and Science Study*, OECD, Paris.

OECD (2004), "ICT Diffusion to Business. Country Report: Norway", DSTI/ICCP/IE(2004)5/FINAL, OECD, Paris.

OECD (2005a), *Governance of Innovation Systems, Synthesis Report – Volume 1*, Organisation for Economic Co-operation and Development, Paris.

OECD (2005b), *Enhancing the Performance of the Services Sector*, OECD, Paris.

OECD (2005c), *Learning a Living, First Results of the Adult Literacy and Life Skills Survey*, Statistics Canada and Organisation for Economic Co-operation and Development, Ottawa and Paris.

OECD (2006), *Going for Growth 2006*, OECD, Paris.

OECD (2007a), *OECD Economic Surveys: Norway, 2007*, OECD, Paris.

OECD (2007b), "The Territorial Review of Norway", GOV/TDPC(2007)9, OECD, Paris.

OECD (2007c), *Education at a Glance 2007. OECD Indicators,* OECD, Paris.

OECD (2007d), *Economic Outlook 82*, 2007, No. 2, OECD, Paris.

OECD (2007e), *Science, Technology and Industry Scoreboard* OECD, Paris.

Oulton, N. (1998), "Competition and the Dispersion of Labour Productivity amongst UK Companies", *Oxford Economic Papers*, No. 50, pp. 23-28.

RCN Research Council of Norway, (2005), *Det norske forsknings- og innovasjonssystemet – statistikk og indikatorer 2005*, Oslo.

RCN (2006), *Annual Report 2006*, Oslo

Remøe, S.-O., M. Fraas, A. Kaloudis, Å. Mariussen, R. Røste, F. Orstavik and S. Aanstad (2004), "Governance of the Norwegian Innovation Policy System, Contribution to the OECD MONIT project", NIFU STEP, Oslo.

Siggel, E. (2006), "International Competitiveness and Comparative Advantage: A Survey and a Proposal for Measurement", *Journal of Industry, Competition and Trade*, 6, pp. 137-159.

Smith, K. (1996), "The Systems Challenge to Innovation Policy", in W. Polt and B. Weber (eds.), *Industrie und Glück. Paradigmenwechsel in der Industrie und Technologiepolitik*, Vienna.

SSB (Norwegian Statistical Bureau) (2006), *Statistical Yearbook 2006*, SSB, Oslo.

St. Meld Nr 7 (1977-1978), Small and Medium-Sized Industrial Firms (Små og mellomstore industribedrifter).

St. Meld. Nr. 28 (1988-1989), On Research (Om forskning).

St. Meld. Nr. 29, 1988-89, Policies for regional development (Politikk for regional utvikling).

St. Meld. Nr. 39 (1988-1999), Research at the start of a new era (Forskning ved et tidsskille).

St. Meld. Nr 25 (2004-2005), On regional policy (Om regionalpolitikken).

OECD PUBLICATIONS, 2, rue André-Pascal, 75775 PARIS CEDEX 16
PRINTED IN FRANCE
(92 2008 01 1 P) ISBN-978-92-64-04373-2 – No. 56277 2008